Glasgow

MAPPING THE CITY

Glasgow

MAPPING THE CITY

John Moore

In association with the National Library of Scotland
and University of Glasgow Library

First published in Great Britain in 2015 by
Birlinn Ltd
West Newington House
10 Newington Road
Edinburgh
EH9 1QS

Reprinted 2017

www.birlinn.co.uk

in association with the National Library of Scotland
and University of Glasgow Library

ISBN: 978 1 78027 319 8

British Library Cataloguing-in-Publication Data
A catalogue record for this book is available on
request from the British Library

Designed and typeset by Mark Blackadder

Printed and bound by PNB Print Ltd, Latvia

Contents

CONTENTS

CONTENTS

Sketch Plan
of
THE CITY OF GLASGOW,
compiled in illustration
of
"Protocols of the Town Clerks
of Glasgow", 1547, et seq.
1894.

St Roche's Chapel
and Cemetery

Glasgwfield

Garngadhill

Town Mill

Lands   of   Cowcaldanes

Subchanters
Croft.

Wester   Craigs

Burn

Glasgw   called   Glasgw

Crowmoule   Lone

Church   of   St Roche

Vicars Yairds

Metropolitan
Church

Provansyde   Lands

Lands of
Swans Yett

Port
Yaird
Castle

Doghillok

Stabligene

St Nicholas
Hospital

Church

Quadrivium
or
Wyndeheid

Medowflat
Lands

Crubbis
Croft

Lone

Deyne syde

Drygait

Malyndinor

Wester

Ramshorne   Lands

Well

Vennel

Eist Brig

Palzait
or
Pallioune
Croft

Commone   Lone

Sculhous

College and
Yard

Blakfriars Kirk,
Cemetery, and
Yard.

Burn

Dowhill

Lang   Croft

West Port

Burn   called   Poldrait

St Tenus
Chapel
and Cemetery

St Tenus   Gait   or

Tronegait

Old Vennel

Muir

Brwmelaw
Croft

St Tenus   Croft

Mercat Cross

Eist Port.

Vennel

Gallowgait

Little St Kentigern's Kirk,
and Yaird.

Gallow

St Tenus Well

St Marys Ch.
and Cemetery

Burrowfield

Stockwalde

Mwtland   Croft

Egliscemeis
Croft

Cropnestock   Croft

Glasgw   Green

Bridgeat

South Port

Burn

Lands

of

Gorbaldis

Brig of Clyd

WATER   OF   CLYD

Linnings
Haugh

Peitbog

Crooks   called   of

Kinclaith

Mildm

Camlawchty

Brwmelands

and

Brigend

NOTE_ It is supposed that in 1547 there
were no buildings on the North side of
Tronegait, West of the West Port.
   Those shown on the Plan were of a
later period.

ROBERT GARDNER & CO. LITHOGRAPHERS GLASGOW

Introduction

This book is about Glasgow, Scotland's largest city, and its maps. Which Glasgow? As the city's current promotional literature emphasises, 'People Make Glasgow' and, in many respects, there are as many Glasgows, real or perceived, as there are citizens. This is no one-way process and the reverse observation that 'Glasgow Makes People' may be seen in the resourcefulness, vibrancy and uniquely individual character of the metropolis. Such a relationship is dynamic and the city of one generation can be viewed in a markedly different way from that of the next. This is, of course, true for any urban area but certain key elements have pivotal roles in any narrative. Geographical location and immediate topography can determine how each settlement develops. History, the character, beliefs and vision of leading citizens and the distribution of wealth also stamp their mark on a city's identity.

Human activity in the Glasgow area has a long pedigree, as evinced by the number of Neolithic canoes discovered since 1780 in localities as widespread as St Enoch Square, Drygate and Glasgow Cross. However, a more definite history stems from the establishment of an ecclesiastical settlement on the site of St Ninian's cell at the end of the sixth century. While it grew into an important religious centre and its cathedral became the finest of its kind in Scotland, Glasgow had little in the way of political importance. It has been estimated that by 1556 its population was about 4,500. Strategically, it had no major significance and this is reflected in the dearth of local maps before the mid-eighteenth century.

Those depictions of late-medieval Glasgow are exercises in historical reconstruction, mostly compiled at the end of the nineteenth century to illustrate works relating to the city's early history. Although they are based on a later understanding of a possible reality, in some ways, they are no more or less selective than any of the maps discussed in these pages. In a time when Scotland treads a rather uneasy path through a maze of questionable heritage artifacts, influenced by what may be described as the 'Braveheart phenomenon' or the minor industry in reproducing work in the style of Charles Rennie Mackintosh, criticism of such reconstructions should be circumspect.

Over the last century, perceptions have been strongly influenced, first, by Glasgow's role as the second city of the British Empire, and then by a post-industrial revision following its subsequent economic decline. For several decades, Glasgow hid its light as a city of note. Thankfully, it has risen to the

---

Alexander McDonald, *Sketch Plan of the City of Glasgow, compiled in illustration of 'Protocols of the Town Clerks of Glasgow'* (1894)

challenge and not only re-discovered a civic architectural richness in which it can be justifiably proud (as seen in the restoration of the Merchant City) but also promoted itself successfully as a popular cultural, entertainment and sporting venue.

Maps reflect more than the changes in society. Their depictions are also influenced by trends in both artistic style and areas of focus. The same can be said for the features affecting the city's history. While some elements appear immutable, the passage of time affects their influence. For Glasgow, this can be most clearly seen during the twentieth century. In 1921, J.W. Gregory contributed an essay on geographical history to the Royal Scottish Geographical Society's *The City of Glasgow: its origin, growth and development*. He highlighted certain advantages of site–proximity to the Clyde estuary, a position on the western edge of the Lanarkshire coal and iron fields, convenience as a business centre for southern Scotland and its situation on the easiest direct route from the Highlands to the north of England.

Glasgow's early growth may have been the result of its position as the lowest fording point of the Clyde but, being fourteen miles from its mouth and that commodious harbour, it needed the ingenuity of the country's leading engineers to turn it into a significant port. However, the narrowness of the channel up to Glasgow and the city's location away from the major centres of Britain's population have prevented it from competing successfully in an era when ships have grown larger and the bulk of many commodities travel by road. As a result, Glasgow still has to come to terms with a maritime heritage which appears to have little to contribute to the contemporary city. The Clyde is still there but Glasgow has yet to integrate it fully back into its daily life.

Additionally, in a time when renewable energy is of major concern and heavy industry has all but disappeared from the economic base, proximity to what are now exhausted mineral supplies has relatively little value. Although the city retains a regional importance as a trade and commercial centre, its west coast location has little advantage when commerce is increasingly focussed on Europe. Of greater importance today is Glasgow's positioning itself on the world stage through its successful marketing campaigns. Glasgow cannot be seen in complete exclusion to its hinterland and this is particularly true when it comes to the Clyde. Due recognition of this close relationship is made in the choice of certain maps. This wider area has had a strong influence in the city's development and in the production of maps, as different administrations tackle the planning and boundary issues of local government.

Despite the relatively short period covered by these maps, the range reflects some of the highs and lows in the history of the city. More than that, they give some insight into the tensions, motives and needs behind their creation. They tell only part of the Glasgow story and there are other parts which were destined never to come to fruition, as can be witnessed by some of the nineteenth-century street proposals or the even more fundamental changes put forward in the immediate post-war period.

I have attempted to increase awareness of the rich map resources held not only in the major Edinburgh repositories, in particular the National Library of Scotland and the National Records of Scotland, but also in various collections in the city itself. A significant number of lesser-known items have been included to encourage further study. Like any work of this kind, the book is selective and reflects the author's own prejudices, interests and knowledge. Readers will come to their own conclusions about the sufficient merit of the text in honouring a very unique, inspiring and fascinating city.

Acknowledgements

I owe a considerable debt of gratitude to Hugh Andrew and his staff at Birlinn for the commissioning of this work.

Without doubt, this book could never have been completed in time without the assistance of Chris Fleet, Senior Map Curator at the National Library of Scotland. His work in the digitisation and geo-referencing of historic mapping, as well as providing online access to more than 100,000 maps through the Library's website has provided an invaluable tool for further research. On top of this, Chris has an enviable record of research in the history of Scots cartography himself and has been ever generous in sharing his extensive knowledge.

I also want to record my thanks for their support to Elaine Anderson, Brian Bell, Sonny Maley, Lesley Richmond, Niki Russell (University of Glasgow), Dr Irene O'Brien and her staff (Glasgow City Archives), Jane Brown (National Records of Scotland), Colin Carter-Campbell, Alasdair Gray, Carol Parry (Royal College of Physicians and Surgeons of Glasgow), John MacKenzie (Royal Faculty of Procurators in Glasgow) and Chris O'Connell (Scottish Canals).

Over the years, many other researchers have encouraged and shared ideas with me. I hope that this work, in a small way, recognises their contributions. Any errors in this text, however, are of my making and I take full responsibility for them.

Dates
The main date preceding the individual entries usually relates to the date of survey, compilation or publication, where known. These dates are also signified in bold, e.g. (**1820**) within the text as cross-references to other entries in the book. The caption to each image provides the title, bibliographic details and the date of publication.

In addition to the generous support of the National Library of Scotland, several other institutions and individuals have similarly granted permission to reproduce images of material held within their collections. I have pleasure in recording these below:

Glasgow City Archives (p.115, 116 17, 235); Alasdair Gray and Strathclyde Partnership for Transport (p.264); British Library (p.12); National Records of Scotland (p.32 (RHP 5974); p.56 (RHP44858); p.77 (RHP5304); p.88–9 (RHP 14283); p.128–9 (RHP95045); Royal College of Physicians and Surgeons of Glasgow (p.134); Royal Faculty of Procurators in Glasgow (p.169); Scottish Canals (pp.44–5); Scottish Enterprise (p.262); University of Glasgow Archive Services (p.182, 184); University of Glasgow Library: University Librarian (p.x, 81, 97, 106, 142, 143, 147, 155, 161, 165, 172, 180, 185, 189, 201, 240, 243, 247, 248, 251, 258 9); Special Collections (p.20, 23, 40, 41, 53, 54, 70, 72, 73, 92, 93, 120, 135, 151, 177, 197, 218, 226, 239); and the permission of private individuals (p.18, 57, 60–1, 62–3, 264)

GLASGVA

Newtou
Gulcadens
Blythes-Wood
Stoberos
Barrowfeld
Gorbals
Kilpatrick
RVGLAN
Spittell
Clyds
Foularton
Kennmort
Conflens
Holhel
Dunpelder
Kersock
Kirckie
Hiltou
Christo.
Provand L.
Rachesy
Carmunock

1596

The first map of Clydesdale

The surviving manuscript maps drawn by Timothy Pont, now held in the National Library of Scotland, are the earliest depictions of many parts of Scotland. Glasgow is no exception to this but, uniquely, this first plan of the burgh, albeit a small sketch on a much larger map, appears on his only manuscript carrying a date: 'Sept. et Octob: 1596 Descripta'. Pont was a son of Robert Pont, a leading figure of the Scottish Reformation, and his work has been much studied in recent years. It is now thought that on leaving St Andrews University in 1583 he soon began work on recording the topography of Scotland. While there is no certainty about the reasons behind this, it was part of a trend across Europe towards a greater understanding and awareness of territory by both monarchs and their administrators.

Those maps attributed directly to Pont contain more than 20,000 place-names and the manuscript which includes Glasgow is one of the most impressive records of the late medieval Scottish landscape. Nearly 1,400 places in Clydesdale are identified, with a heavy bias towards human features, reflecting the population pattern of the day. Pont was not solely a surveyor but, following the style of the time, more of a chorographer, describing the landscape and recording important information on genealogy, local gentry and other such information which reflected a stable, ordered society. This Clydesdale map is a more finished draft, based on earlier fieldwork, and suggests that Pont may have completed his surveying by the summer of 1596.

Glasgow is clearly identified as a linear settlement running north from a bridge crossing the Clyde and terminating at the High Kirk. A careful scrutiny would suggest that Pont identifies the slant of Bridgegate running from this bridge to the Saltmarket. The market cross is clearly located at the junction of the Saltmarket-High Street line and the lower of two streets which cross this backbone of the town. A further

Timothy Pont, Detail from [*Glasgow and the county of Lanark, Pont 34*] (1596)

Timothy Pont, Detail from [*Renfrewshire, Pont 33*] (c.1596) showing a different orientation

early plan exists. Scot, however, was sympathetic to Glasgow, for, in the aftermath of the disastrous 1652 fire, he endowed a fund with land from his Fife estates to support the training of six poor apprentices. Joan Blaeu, one of Europe's leading cartographers, had begun work already on his great atlas, *Theatrum Orbis Terrarum, sive Atlas Novus*, and used the manuscripts as the basis for the fifth volume of this landmark work, which was published eventually in 1654.

The Clydesdale manuscript was to be copied closely by the Dutch engravers but, as Pont had died around 1614, there are several errors in the transcription of information from the original. In comparison with Pont's work, the atlas version of Glasgow on the printed map, *Glottiana Praefectura Inferior*, has lost the significance of the cathedral dominating the north of the town and there is no indication of the market cross. On the other hand, the name 'The Mills' on the east bank of the Molendinar is clearly visible, indicating its significance as a source of power. One additional feature is a route running north towards Kirkintilloch.

The accompanying text in the atlas is based on Camden's *Britannia* and sings the praises of the city: 'the most famous town of merchandise in this tract: for pleasant site, and apple trees, and other like fruit trees much commended, having also a verie faire bridge supported with eight arches'. It also includes lines from *Poemata Omnia* by Arthur Johnston, one of the major Scots Latin poets of the early seventeenth century:

Glasgow, you hold out your head among allied cities,
And the great globe has nothing more beautiful than you.
Under the western sun Zephyr's breeze cools you,
And you fear neither frost's cold nor Dog's face.
The Clyde girding your side is purer than all amber,
Here you rule a thousand sails with your power.

Glasgow is also included on the neighbouring Renfrewshire manuscript (Pont 33). Here, the town is presented as a linear settlement running parallel with the Clyde, but identifying the cathedral as the most dominant building. Again, there is no indication of Gorbals south of the river but a road runs from

road runs north-west from the river to join the Trongate. In addition, the grounds to the east of the College and Blackfriars lands are defined, as are two other ecclesiastical buildings, presumably the churches of the Blackfriars and St Mary and St Anne (later the College chapel and Tron Kirk). The Molendinar burn marks the town's eastern boundary but the map shows little of a settlement south of the river. Recognisable names in the surrounding countryside include *Barrowfield, Carntyn, Coucadens, Rudry, Camlachie, Stobcros* and *Blythswood*.

Largely through the intervention of Sir John Scot, Director of Chancery and a Privy Councillor, who had prepared a volume of Latin poems for the Dutch publisher Blaeu, these maps were sent to Amsterdam. In 1641, Scot had obtained royal backing to publish Pont's work and gained support from the church's General Assembly for the compilation of written accounts to accompany it. That June, the Glasgow Council employed James Colquhoune to draw a portrait of the town 'to be sent to Holland'. Unfortunately, no trace of such an

Timothy Pont/Joan Blaeu, Glasgow from *Glottiana Praefectura Inferior cum Baronia Glascuensi* from *Theatrum Orbis Terrarum sive Atlas Novus* (1654)

Timothy Pont/Joan Blaeu, Glasgow from *Praefectura Renfroana vulgo dicta baronia* from *Theatrum Orbis Terrarum sive Atlas Novus* (1654) with a noticeably different street pattern

the bridge to Paisley. Interestingly, the image of Glasgow which is shown on the resulting atlas map, *Praefectura Renfroana*, which directly attributes Pont as its author, is closer to that of the Clydesdale depiction, with the same mills marked and a greater sense of the north-south orientation of the town, albeit with a double street pattern. It should be remembered that these maps were published uncoloured and the ability of the subsequent colourist can have an impact on how a settlement is seen by the level of care with which the colouring was applied.

1693

A view of the City from Wester Craigs

Nearly a century elapsed before the next portrayal of Glasgow's townscape was to appear, in John Slezer's *Theatrum Scotiae*, first published in 1693. The subtitle of the book informs the reader that it contained 'prospects of their Majesties castles and palaces, together with those of the most considerable towns and colleges'. It provided, for the first time, a comprehensive view of the nation's most significant burghs and buildings. Although described as 'high German', Slezer was an army officer and an able military draughtsman who had the acquaintance of several of the Scots nobility. In September 1671, he was appointed as Chief Engineer for Scotland. His duties included a review of the condition of the country's fortifications and, during his tour of inspection, he formulated a plan to produce several volumes of views.

It is most likely that a camera obscura was used in the production of his drawings, allowing the projection of a scene onto a flat surface. This prospect of the town from the north-east was clearly taken from the high ground of Wester Craigs, to the east of the cathedral. It is this building which dominates both the town and the plate. Slezer provides a valuable image of the irregular western towers of the High Kirk which were demolished in the 1840s in the expectation of their reconstruction to a more acceptable design. Of possibly greater importance to the history of the city, he shows the central keep, great tower, gatehouse and curtain wall of the Bishop's Castle, all in an apparently good state of repair. Although counter to the accepted version of them falling into ruin after 1688, more recent research confirms that efforts to preserve the buildings continued until, at least, the 1740s. Garrisoned by French troops during the Regency of Mary of Guise in the 1550s, the castle was to have a chequered existence until its last clerical occupant, Archbishop John Paterson, left on the abolition of episcopacy in July 1689. The castle then became crown property. Three years after this view was published, Lord

John Slezer, *The Prospect of ye Town of Glasgow from ye North East* from *Theatrum Scotiae* (1693)

Keep, tower, gatehouse and curtain wall of Bishop's Castle

The city's several spires and steeples located in the lower town

Cathcart was granted liberty to occupy the house but the buildings were finally demolished completely in 1789 to make way for the construction of Glasgow Royal Infirmary.

Elsewhere in this prospect are the spires of the Old College, Blackfriars Church, the Tolbooth, the Tron Kirk and the steeple of the Merchants' Hospital on the Bridgegate. In the accompanying text, which was written by Sir Robert Sibbald in his capacity as Geographer Royal and possibly part of his own unsuccessful scheme to publish a Scottish atlas, 'the number and stateliness' of the city's private and public buildings is praised. Slezer clearly attempted to match this in his drawings. He delineated the considerable walls surrounding the College grounds carefully and, in the foreground, a small bridge crossing the Molendinar Burn and

leading into Drygate. However, the rising ground to the south-east of the town seems entirely fanciful.

This is one of three views which Slezer prepared to illustrate the city; the others being a prospect from the south, which shows the old bridge, along with its port, and one of the College buildings. The southern prospect was drawn at too great a distance from the town to provide much detail but it does emphasise how flat the land beside the Clyde was and how easily it could be flooded. It should be noted that, while Sibbald's description emphasises Glasgow's status as 'the most famous Empory of all the west of Scotland', Slezer makes no effort to indicate either trade or industry in the town, preferring to suggest its setting in a markedly sylvan scene. Sibbald also repeats Johnston's poem of praise of Glasgow which

The bridge and Water Port from Slezer's *Prospect of ye Town of Glasgow from ye South* from *Theatrum Scotiae* (1693)

appeared in the Blaeu atlas. This, combined with the dominance of the cathedral on this plate, could suggest that Slezer was prone to a focus on the past rather than a dynamic future.

While Edinburgh had a flourishing publishing trade by this time, the drawings for this work were sent to be engraved in London and Holland. In addition, the completed book was published in London. As stated, it was Slezer's intention to produce further works covering country houses and other significant buildings but, unfortunately, the *Theatrum* was not a commercial success and he spent much of the rest of his life in penury. No other volumes appeared. In addition, he had translated Sibbald's Latin text into English without consulting him or acknowledging his authorship, leading to a breach

between them. Slezer was not alone in this, for a similar situation developed between Sibbald and the cartographer John Adair. Despite Adair's extensive work surveying Scotland, he never produced maps which cover the Glasgow area in any detail and he too struggled financially. It was a time of high intention but limited funding and Adair and Slezer had to compete for the monies voted to them by parliament from a levy on the tonnage of ships entering Scottish ports. Slezer died in 1717, having moved to the debtors' sanctuary of the Holyrood Abbey grounds and never receiving the support promised him. Despite the lack of sales, at least another four editions of the book were published up to 1814, often without text but using the original plates.

LOCH

Skipness

Scale I.

Ramsay
Loch

The Cary
Kilfinea
Prou
Loch Gair

Brodick Castle

Aird
Glenderavel
Castel Lathlan
Loch Fine
Iner Lockan
Penymoor
Letir

Inch-marnock
Etrick Bay
Buttock
Calon
Loch Ridden
Killbryde
Ine

Keanes
Loch
Bute
Rothesay
Anernie
Loch Heck
Strachur
Crogou

Kyles of Bute
Loch
St. Cathrines

The Mouth of Clyde River
Mount Stenray
Scenley
Ascoy
Ardine
Rowan
Inelon
Dunnon
Ardintenie
Kilmun
Port Dornick
Carrick
Ardkinglas

Little Comra
16   16   16
Holly Lock
Loch Long
Loch Goile
Loch Goils head

Great Comray
11   10
Grichen
Loch Goile

Pen Corse
16   16   24
Clock Points
Weems Point
Ardrou
Baran
Roseneath
Cumsleo
Clachan
Rachen
Port Chapol
Ardgarton

Hunter ston
Knoc Skirmorly
Kelly
Innerkip
Gourock
Ross
Gare
Loch
Ron
Ardinchaplen
Ben Veen
Loch

Fairly
Kelburn
Largs
Greenock
Glen Douglas
Ben Tarbet
Tarbet

Killbryde
Cunningham
R
Cartsdyke
Milardmore
Glen Froon
Luss
Loch

Horse I.
Ardrossan
EN
Port Glasgon
Newark
Cardros
Rosdoe
Inch Mirin
Lomond

Saltcoats
Killbirny Loch
Lochunoch
Kilmalcom
Finlayston
Ardoch
Bonhill
Loch
Ben Lomond

Killnining
Castle Temple
Kilelan
Leveniside
Bal-maha

Irvin
Beeth
Houston
Dunbarton
Buchanan Castle
Quarick R.

Division
Kilbarchan
Bishopton
Dunglas
Kilpatrick

Broon
Johnston
Erskine
B
Damure
Killern

Kilmarnock
FR
Niliston
Paislay
Northbar
Inchgreen
Scotston
Jordanhill
Forth River
Stir

R
EW
Hatket
Renfield
Renfren
Barns
RI

Shire
Croxston
Govan
Parlick
TAN
Gargunnock

Mearns
Eastwood
Kelvin

Cathcart
Gorbes
GLASGOW
Calder
Shire

Shanfield
Barronsfield
Stirling

Rugland

1731

'Published for the Good of the Publick': Adair's map of the Clyde

Towards the end of the seventeenth century, one name in particular stands out as the leading Scots surveyor of his day. This was John Adair (1660–1718), arguably one of the most enigmatic of Scottish cartographers and variously described as 'Hydrographer Royal', 'Geographer for Scotland' or 'The Queens's Geographer in these Parts'. He first came to prominence as a surveyor of the eastern counties of Scotland but is also remembered for his charting of the coastal waters. By the mid-1680s, he had already completed charts of the Forth and the Clyde but it was only after the Scottish Parliament passed a tunnage act in 1686, specifically to defray his expenses, that Adair's attention began to focus on this hydrographical work. Overseas trade was growing and, as the act states, his surveying was considered 'most necessary for navigation and may prevent several shipwrecks'.

It is not certain what prompted Adair to embark on such a scheme, particularly at a time when Captain Greenvile Collins, with the support of the Admiralty and Trinity House, had already begun charting the coasts of Britain. Collins was engaged in this work between 1681 and 1688 and the resulting maps were published in 1693 in *Great Britain's Coasting Pilot*. It includes eight detailed charts of the east coast from Berwick to the Orkney and Shetland Islands but no coverage at all of the west coast. It is possible that the Scottish authorities wanted something specifically useful for Scottish mariners and employed Adair for this purpose. Despite criticism of the Collins charts, the *Pilot* was republished in 1723, by Mount and Page, and a further nineteen editions were issued up until 1792. This contrasts sharply with Adair's own atlas, *The Description of the Sea-Coast and Islands of Scotland,* which appeared in 1703. Although this was to be the first in a projected series designed to cover all Scottish waters, this was the only edition to be realised. It similarly concentrates on the major features of the eastern seaboard south from Aberdeen,

John Adair/Richard Cooper, *A new and exact map of the River Clyde* (1731)

The Tail of the Bank and the shoals upriver

still the most important areas of mercantile interest.

Like his contemporary John Slezer (1693), Adair was plagued by a lack of guaranteed financial support, exacerbated by a difficult business relationship with Sir Robert Sibbald, whose own unsuccessful project to produce a new atlas of Scotland seems to be behind much of the geographical work of the day. Relatively few of Adair's charts have survived and this has hindered a proper assessment of his hydrography. In 1713, Adair applied to the Admiralty for support and appended a list of his unpublished surveys, which included four charts of the Clyde estuary 'in the greatest forwardness', requiring little time or expense to finish them. These appear to be the result of two expeditions to the firth in 1696–97. Certainly, there are no printed charts by Adair covering the west or south coasts. What have survived are three versions of a manuscript of the estuary between Irvine and Ayr, which

can be dated to 1686. These give the impression of being only partly complete – for example, while Adair details the extensive shoal water south-west of Fairlie, the soundings marked in the Ayr Roads are not replicated for the Largs Channel. One specific item from Adair's submission, 'A general Map, of all the west and south coast, from Carlisle round to Glasgow, drawn out ... but can not be finished till the Isles of Arran, Bute ... Kintyre and part of the coast of Ireland be surveyed and joyned to it' may have been the base for this 'new and exact map of the River Clyde'.

Richard Cooper (1701–64) moved from London to Edinburgh in about 1725 and quickly established a successful business, based largely on the production of portraits and book illustrations. Cooper's entrepreneurial skills developed and he moved into interior design for both institutions and the nobility. However, it was in the field of engraving that he

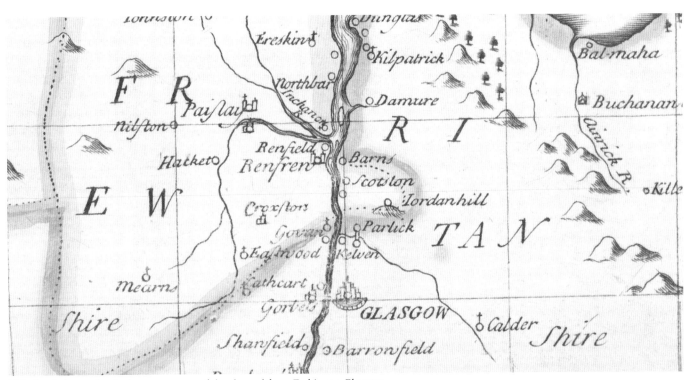

This detail shows clearly the narrowness of the channel from Erskine to Glasgow

dominated, being responsible for most of the maps and prints of any consequence then produced in Scotland. Apparently, Cooper purchased Adair's copper plates from the cartographer's widow at a bargain price and, in 1730, produced a map of the Forth based on the surviving charts. The Forth map was followed in 1731 by this Clyde depiction and, between 1735 and 1737, county maps of the three Lothians.

The first thing about this chart which catches the reader's attention is its orientation, where west is at the top, allowing the map to be centred on Argyllshire. Given that it is dedicated by the publisher, George Scott, to the 2nd Duke of Argyll, this may be a little less surprising. While the sweep of the bays at Irvine and Ayr are exaggerated, the chart clearly identifies the main channel to the west of the Cumbraes, as well as the sandbanks and shoals upstream. This is particularly noticeable east of Greenock where the image emphasises just how restricted the navigation was. What is significant is the variance between the coastline as indicated on Adair's manuscripts and Cooper's engraving, which has an obviously different shape to Arran and completely ignores Troon Point and Lady Isle. Apart from these discrepancies, it should be remembered that the information shown here may have been between thirty and forty years old.

Undoubtedly, Adair's main weakness was his marked inability to see his surveys through to final publication. His money problems certainly did not help but the historical evidence suggests that, regardless of this, he was either over-optimistic or guilty of an inability to prioritise his work. Despite his obvious talent, exemplified by the quality of his draughtsmanship and the greater accuracy with which those manuscripts are drawn, his surviving work only serves to underline what might have been.

1753-54

The Military Survey of Scotland

This second manuscript representation of the city comes from the Military Survey of Scotland, begun in 1747 after the Jacobite Rebellion ended on the battlefield of Culloden. The Hanoverian military establishment had been greatly embarrassed by its lack of proper intelligence of the formidable topography of the Highlands and, following a suggestion from Lieutenant-General David Watson, then Deputy Quarter-Master General of the army in Scotland, it was resolved to correct this state of affairs. While it was largely Watson's idea, the 'Great Map', as it was called at the time, was directed by a civilian, William Roy, who would subsequently acquire fame as 'the founder of the Ordnance Survey'. Both men were Scots, Roy being the son of the factor to the Hamiltons of Hallcraig, near Carluke, who had been introduced to Watson through connections with the Dundas family of Arniston.

Initially, Roy was employed as the only trained surveyor, assisted by a small team but, in March 1750, the establishment was increased and eventually numbered about 60, comprising several surveying parties and three draughtsmen. One of these was Paul Sandby, the noted watercolour artist. By 1752, mapping of the Highlands was completed and, at this point, it was decided to extend the operations to cover southern Scotland, with financial stability ensured by funding from both Watson and members of the Dundas family. This southern mapping worked north from the English border using two parties, the western under Roy himself. By 1753, a survey team was known to be working in southern Lanarkshire. The plan of Glasgow, therefore, is most likely to have been the work of the man whose name is so closely associated with the whole enterprise, possibly prepared in the summer of either 1753 or 1754.

The surveyors used simple instruments – chains and circumferentors, which were basic compasses which could measure horizontal angles – with an emphasis on speed and

William Roy, Military Survey of Scotland (1753–54)

13

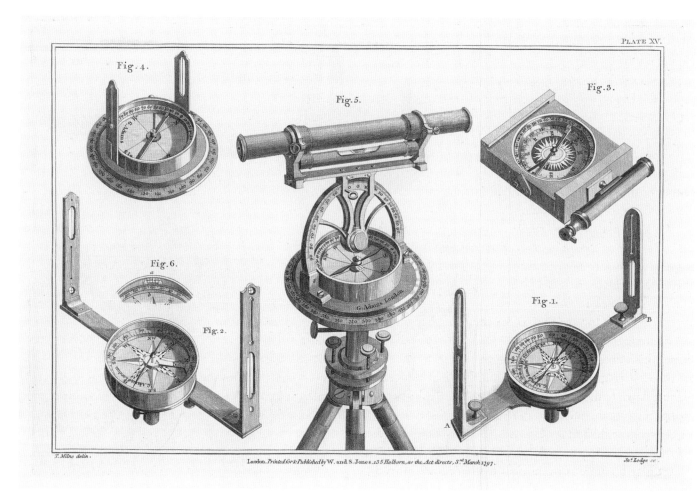

ABOVE. Examples of contemporary surveying instruments, including a circumferentor (fig.4) from George Adams, *Geometrical and Graphical Essays* (1791)

OPPOSITE. This extract emphasises the Survey's attention to street pattern but also the lack of place names

economy, and based on measured valley traverses. Surrounding details were filled in by sketch or copied from existing maps. Fieldwork was adjusted during the winter months in the drawing office at Edinburgh Castle. Every year, the work was inspected and approved by the Duke of Cumberland. The map, now in the British Library, covers only mainland Scotland, at a uniform scale of 1000 yards to an inch, but, nonetheless, represents an outstanding achievement by contemporary standards. It shows a standardisation of style and use of colour which was then being introduced to military cartography, with the choice of red for buildings, green for wooded areas and yellow for cultivated land. No fair copy for southern Scotland was made, the surviving sheets being working maps with a less finished style but, on its completion, Sandby produced a reduction of the whole survey at 12,000 feet to an inch. It should also be remembered that the whole map was oriented to magnetic north, which at this date was 19 degrees west.

Although it is known that pre-existing town plans were used in certain areas, there is no evidence to suggest alternative sources for the layout of Glasgow. By this time, the surveyors were well experienced in the routine of their duties and would have found few problems in delineating the basic urban pattern, particularly in a burgh which was anxious to show its loyalty to the Hanoverian cause. Records, however, show that earlier plans of Glasgow and its royalty were drawn by both John Watt and James Barry in the 1730s and 1740s. As neither of these survive, it is feasible that the military used these as bases for this depiction.

While the plan is at too small a scale for very much detail, it is still a clear and, presumably, accurate picture of the burgh at a time well before any significant expansion took place. The major axes of High Street–Saltmarket and Trongate–Gallowgate are obvious, but with a markedly straight line to Argyle Street. The Cathedral and the North-West Church at the head of Candleriggs are both depicted by a standard cruciform shape which does not reflect their true plan. Also identified are the College grounds and the slanting line of Bridgegate running down to the old bridge. A grey colour wash has been used to indicate the rocky promontory of Wester Craigs in a far clearer way than other eighteenth century city plans but the depiction of the field boundaries of the neighbouring estates may be more conventional than accurate. Already,

Glasgow can be seen to be the centre of a network of roads emanating out in all directions from the Cross. What is quite striking about the map is the relatively small number of place-names in the immediately surrounding countryside but, interestingly, they include *Oakinghous* and *Tennants*. More importantly, the map identifies a significant settlement on the south bank of the Clyde at *Gorbels*.

While there are variations between what is shown on the 'Original Protraction' and the 'Fair Copy' of the northern survey, the whole map remains a most valuable document for the study of Scotland's landscape and rural economy, as well as urban form, routeways and the layout of estates, at a time before the major eighteenth-century agricultural improvements began to have an effect. Whatever its strengths and failings, the completed work was passed to Watson and subsequently entered King George III's library as part of his topographical collection. It was neither published nor exhibited to a wider audience and seems to have played little part in subsequent military strategy in Scotland. It was not seen or used by later surveyors until the following century, when Aaron Arrowsmith consulted it for his own map of Scotland. Of possibly greater importance to Scotland's cartography was Watson's role on the Board for the Annexed Estates in influencing the manner in which they were to be surveyed.

1759

A family affair: the Watt chart of the Clyde

Glasgow's developing mercantile prosperity during the early eighteenth century was hampered by the condition of the River Clyde, particularly upstream from the Tail of the Bank off Greenock. Constant silting and submerged shoals, combined with a very shallow channel, meant that, at low tide, navigation for vessels with any sizeable draught was virtually impossible. Earlier efforts to improve the situation met with little success and the creation of alternative deepwater facilities at Port Glasgow still left problems for the trans-shipment of goods inland. Increasing trade and the poor condition of the connecting roads convinced both merchants and Burgh Council that the most effective long-term solution was to deepen the Clyde. In 1755, John Smeaton, the foremost engineer of his day, was invited to investigate plans for a dam and lock on the river. Accompanied by James Barry, he found the minimum depth of water from Renfrew up to the Broomielaw to be less than eighteen inches. To permit the necessary improvements, the Glasgow magistrates petitioned parliament and, despite strong opposition from other Clyde burghs, the relevant act received Royal Assent in June 1759. It is assumed that this is the background to this chart's production. However, on closer investigation, it poses many questions.

Although it is attributed to John Watt, he had died more than twenty years earlier, in 1737. Furthermore, he is known more as a talented land surveyor, employed by, among others, the Dukes of Montrose and Hamilton. While James Watt confirmed that his uncle's original survey of 1734 extended only to Toward Point and that the rest was added by his father and brother, there is no evidence to indicate that John Watt senior had been involved in recording the river's shoals, offshore rocks and soundings. Despite this, an anonymous surviving chart, complete with scale bar and title and oriented similarly to Cooper's engraving (1731), is likely to be by him.

John Watt, *The River of Clyde* (1759)

17

Anon. *Portrait of John Watt* (n.d.)

In August 1731, John Watt purchased two Clyde plans, presumably copies of this engraving. At about the same time, his finances show a notable growth in money flow, mostly related to shipping transactions. These interests may be behind the manuscript's preparation. Comparison of it with the 1731 map shows marked differences in the coastline, choice and spelling of place-names and, particularly, the indication of a safe channel upriver, where Watt identifies neither rocks nor soundings. Nonetheless, there are sufficient similarities to suggest that this draft formed part of the compilation process for the map finally published by his family.

James Watt's father was a Greenock merchant and burgh treasurer. While it is clear that he was involved in the complex production process behind this map, there is no evidence to suggest that it was prepared as part of Greenock's opposition to the act. It is more likely that his son responded to the consid-

erable interest prevailing in Glasgow at the time. The finished work was first advertised in October 1759 as available from James Watt's workshop within the University, where he worked as a mathematical instrument maker.

The Clyde is an awkward estuary to show on a map, particularly if the intention is to indicate both the approaches to the firth and the course to Glasgow itself. Any chart including both covers a wide area and either needs a very large sheet or is drawn at a scale too small to be useful to mariners. The Watts tried to circumvent this by an unusual combination of two separate sections, consisting of a more detailed illustration of the Clyde from Garroch Head, at the southern tip of Bute, to the turnings of the river east of Glasgow, arranged on two sides of a general map of the south-western approaches to the estuary, from Mull to the coast of Northern Ireland. The title states it was published 'According to the Best Authorities' which suggests that the family had little to do with any detailed survey – a likelihood strengthened by the general lack of any marine features apart from the waters around Islay, which was of particular relevance to James Watt senior's business concerns. In addition, the larger scale depiction of the river provides good detail of the anchorages at Greenock.

What is also noticeable about the sheet is that the two representations do not have a flush margin, and comparison of the area similar to both shows obvious differences, especially in the shape of the Ayrshire coast. James Watt was involved in a subsequent Clyde survey in 1758 but, more importantly, family letters indicate that his role in the production of this map was more editorial, particularly in balancing his father's rather finicky suggested corrections with the practicalities of engraving such a large sheet. It now seems likely that there was a major change of thought in March 1758, following the engraver's advice to add a meridian to the finished map. If it was deemed necessary to indicate degrees of longitude and latitude, the orientation of John Watt's manuscript was a major problem. This may have led to the family discarding the map but not what the map showed, for there are strong similarities between the coastline on it and the engraved chart.

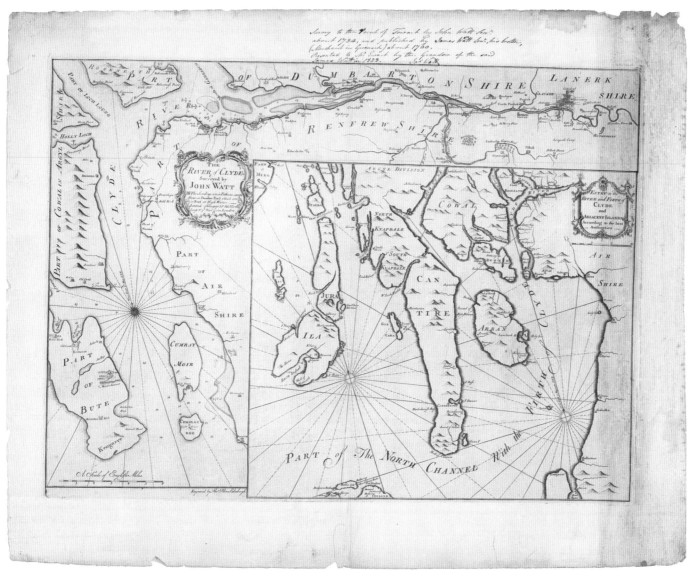

The complete sheet highlights the differences in the depiction of parts of the Clyde estuary between the two charts

The documentary evidence also supplies a picture of Watt's negotiations with the Edinburgh engraver Andrew Bell, but it is Thomas Phinn's name which appears on the published map. Equally illuminating is the fact that the family were charged three times the original price quoted for printing. This was to be Watt's only venture into map publishing in this fashion and, despite the involvement of so many individuals, the lock project itself was abandoned at the end of 1762, leaving the way open for the more effective scouring scheme suggested by John Golborne six years later. Was the map useful to mariners? Certainly, nothing was to replace its coverage until the Admiralty charts a century later.

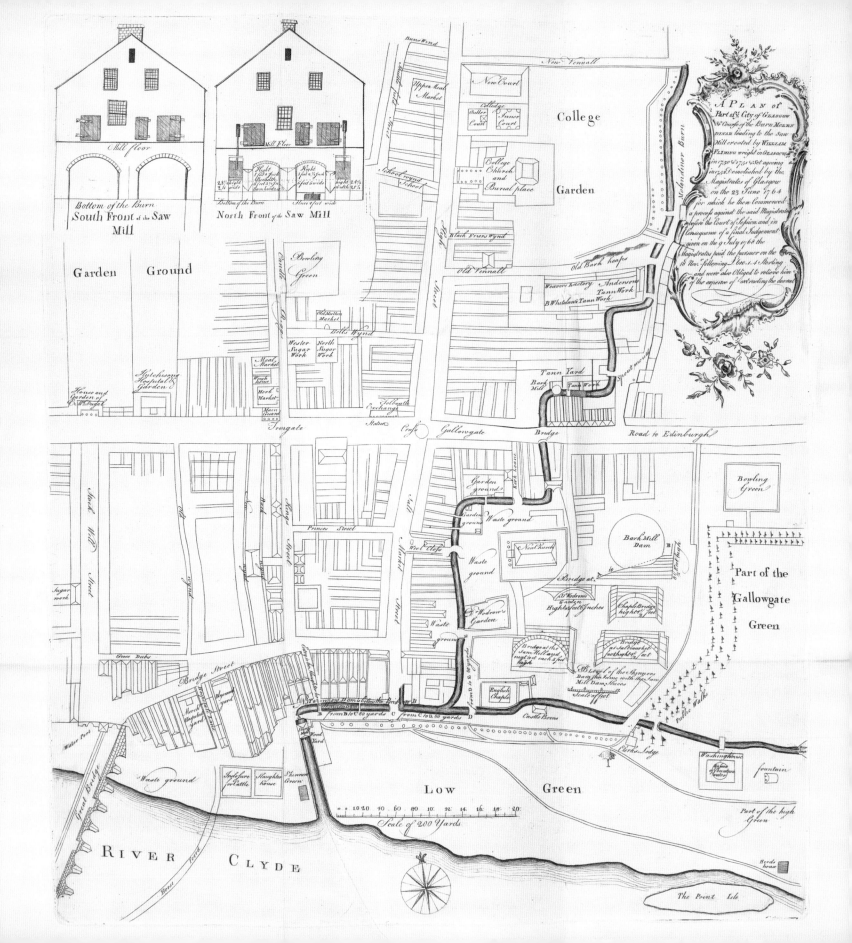

Bottom of the Burn
South Front of the Saw Mill

Mill Floor

Bottom of the Burn   Sluice feet wide
North Front of the Saw Mill

Garden   Ground

Garden Ground

Bowling Green

Candle Riggs

Old Mutton Market

Betts Wynd

Wester Sugar Work   North Sugar Work

Meal Market

Weigh house

Herb Market

Hutchisons Hospital Garden

Main Guard

House and Garden of Wm Inglis

Trongate   Cross   Gallowgate

Burns Wynd

Upper Meal Market

College

School Wynd   School

High Street

College Church and Burial place

Black Friers Wynd

Old Vennall

Old Bark heaps

Weavers Factory   Andersons Tann Work

B. Whitelaws Tann Work

Tann Yard

Bark Mill   Tann Work

Bridge   Road to Edinburgh

New Vennall

New Court

College Outter Court   Inner Court

College Garden

Molendiner Burn

Spout mouth

A PLAN of
Part of ye City of Glasgow
& Course of the Burn Molinx
dinax leading to the Saw
Mill erected by WILLIAM
Plenus wright in Glasgow
in 1750 & 1751 set agoeing
in 1752 Demolished by the
Magistrates of Glasgow
on the 23 June 1764
for which he then commenced
a process against the said Magistrates
before the Court of Session and in
Consequence of a final Judgement
given on the 9 July 1768 the
Magistrates paid the pursuer on the
18 Nov. following £ 600.1.4 Sterling
and were also Obliged to relieve him
of the expence of extracting the decreet

North Mill Street

Old Vennall

Kings Street

Princes Street

Sugar work

Goose Dubs

Bridge Street

Salt

Market Street

Wool Cliss

Waste ground

Waste ground

Waste ground

Kirk Lonne

Garden grounds

Garden grounds   Waste ground

New Church

Dr Wodrow's Garden Highest few Inches

Dr Wodrow's Garden

Bark Mill Dam   B

to   Feet high

Bridge at

Bridge at Saltmercat feet high 9 feet

Bridge at the Saw Mill and inmost each 5 feet high

English Chapel

B level of the old ymers Dam this being with the Saw Mill Dam Slices

Scale of feet

Bowling Green

Part of the Gallowgate Green

Chaple Bridge height 9 feet

Bridge at Saltmercat feet high

Castle Borns

Public Walk

Great Bridge

Waste ground

Inclosure for Cattle   Slaughter house

Shennans Green

Merchants Hospital Yard

Shipmens Yard

Saw Mill Yard

New ymers Dam 1 below the Bridge B

B   from B to C 80 yards   C   from C to D 80 yards   D

from D to E 80 yards   E

Clerks Lodge

Washing house

fountain

Low   Green

0 10 20 40 60 80 10: 12: 14: 16: 18: 20:
Scale of 200 Yards

Part of the high Green

RIVER   CLYDE

Horse Ford

Herds house

The Point Isle

1764

Molendinar Burn sawmill plan

Taken at face value, this engraving is a piece of evidence prepared for a legal case. Although it covers only part of the city, it is probably the earliest surviving printed plan to show Glasgow in greater detail. It was originally prepared as proof for the Burgh Council in an action brought before the Court of Session by William Fleming, a timber merchant with an extensive business in the town. In 1751, he opened a sawmill on the Molendinar Burn but, thirteen years later, the town authorities, claiming that they were facing complaints over the lease, had the mill dam and other machinery summarily demolished. After a protracted case, the court ruled in Fleming's favour in 1768 and the Council was instructed to pay more than £700 in damages and costs.

There had been a sequence of concerns about spates and the flooding of houses close to the burn, as well as the impact that the several dams had on the quality of the water. Fleming believed that the influence of whip-sawers was behind the action taken by the Council. However, when the names of some of the figures involved are investigated, a degree of commercial self-interest may also underlie the demolition. At about the same time, petitions of complaint by local residents had been sent to both the owners of the tannery above Fleming's mill and the magistrates. Andrew Cochran and John Bowman, both partners in the Tannery Company, served as lords provost of Glasgow in the years 1760–62 and 1764–66 respectively. Another member of this business concern was Alexander Spiers, an influential town councillor, so that this company had significant influence in the council chamber. While it was Fleming's dam which was removed, no action was taken with regard to the tanworks.

Centred on Glasgow Cross, the plan shows the four main streets radiating out as far as Stockwell Street, the New Vennel, the Clyde and Gallowgate Green. Although the street frontages tend to be shown as continuous blocks, the plan attempts to

[James Barry], *Plan of part of ye City of Glasgow & Course of the Burn Molendinar* (1764)

detail the burgage plots, wynds and closes in the core area of the burgh and, again, emphasises how important the Molendinar burn was to its early industry. Apart from Fleming's sawmill, a bark mill and three tanworks also made use of its waters. Even more striking is the number of bridges which crossed it in a very short distance. Cross sections of several of these, as well as two elevations of the sawmill itself, fill empty areas of the plan. Three sugar works testify to the growing trade across the Atlantic, while the town's four markets emphasise its importance as a local commercial centre. Several key buildings can be identified, including the College, the recently built churches of St Andrews in the Square and St Andrews-by-the-Green (here identified as 'English chapel'), both set amid gardens, the Main Guard, weigh house and Exchange. The eight arches of the old bridge are carefully shown but of greater historical interest may be the only mapping of both the Water Port gate at the north end of the bridge and the horse ford to the east. Gardens, waste ground and two bowling greens all add to the sense of the rural nature of the town.

In the process of the case, over seventy witnesses were questioned and their depositions provide light on contemporary life in the city and the preparation of this plan. While a surviving manuscript draft exists, it lacks the characteristic block stamping of letters which was a feature of the plans of James Barry, who is most likely to have been the original surveyor. By this date, he had more than thirty years' experience and had been employed frequently by both public institutions and private individuals within the city. Although he had inspected the dam some years before, he stated that he surveyed the Molendinar and Camlachie Burns for the authorities in December 1764, along with the sawmill and dam. This brings in some element of confusion since the dam was removed in the June of that year. Furthermore, several witnesses, including John Riddell, mention measuring dams and levels on the burn. More importantly, the memorial

LEFT. The exceptionally decorative title cartouche

Anon. *Stockwell Bridge (north end) in 1776* (n.d.)

presented by the city magistrates in 1767 indicates that Barry's plan displays not only the course of the burn but also 'the lying of the streets and grounds adjacent thereto'. In his final expense account, Fleming records a half payment of £4-12-6d for the engraving of two plans, the Council paying the other half. In comparison with his solicitor's fee of 15 guineas for 'a very frequent and troublesome correspondence', the relative significance of such a survey can be seen.

The Molendinar was to remain something of a problem in later years until it was covered over along its length in 1877.

An 'Ode to the Molendinar' published in the *College Album* of 1832 may or may not be truly descriptive:

With weariness and pallor fraught
Upon thy clay-built banks I'd lie
And gaze upon thy smoke-veiled sky,
Or watch thine ever-changing hues
And breathe the scents thy waves diffuse
Roll on, with murky billows roll
Juice of the mud-cart and the coal.

A PLAN OF GLASGOW

REFERENCE
a Princes Street
b Baxters Wynd
c Goose Dubs
d Spout Mouth
e Butcher Market
f Veal Market
g Grammar School Wynd
h Stock Friars Wynd
k Dry Gate
l Bishops Palace
m Ratton Row
n Gray Friars Gate
o Dean Side
p New Vennal
q Buns Wynd

High Kirk

College

College Garden

Dove hill

Observatory

Dove hill

Tan Work

Tan Work

Edinr. Road

WestPort   Trongate   Guard

Gallow Gate

Jamaica Street

Rope Walk

Broomie Law

old Green

Ducat Green

Timber Yard

Bridge Street

Slaughter house

Calton

THIS AREA
LAIDE OUT FOR
BUILDING

Laigh Green

Gallowgate Green

Washing house

RIVER CLYDE

GORBELS
Pasley road

Hords house

35

40

45

1   5   10

1773

'Both useful and entertaining to the purchasers': Ross's map of Lanarkshire

The 1770s witnessed something of a flourishing of urban cartography in Scotland, with several burghs appearing in printed map form for the first time. John Ainslie, who was to become the most significant map-maker of his day, produced a four-sheet plan of his home town of Jedburgh in 1770 and went on to include inset plans of Cupar and St Andrews on his county map of Fife and Kinross. Additionally, Andrew and Mostyn Armstrong published similar plans of Ayr, Edinburgh, Greenlaw, Haddington, Linlithgow and Peebles on their maps of Ayrshire, Berwick, the Lothians and Tweeddale during the decade. In other words, this first surviving engraved plan showing the whole of Glasgow was not an exceptional feature of the period, since it lies in the south-west corner of Charles Ross's map of the county of Lanark, seemingly as if on a separate sheet nailed to the larger work.

The plan itself is at a scale sufficiently large to indicate Glasgow's most important buildings, although many are shown solely as blocks without any naming. It covers the town from Calton to the Broomielaw and from the Cathedral across the Clyde to the village of Gorbals. A table of references is added, listing several of the lesser wynds and streets and two of the burgh's markets, emphasising the town's continuing importance as a local centre. Although Glasgow's population was less than 32,000 by the mid-decade, overall we are presented with a picture of a city on the move. Already, a westward spread of the urban pattern can be discerned in the new streets starting to appear running off Argyle Street. These were to become the preferred addresses for a mercantile elite which was already beginning to enjoy new levels of a prosperity founded on overseas trade. The names of some of these streets underline the basis of this wealth. In particular, Jamaica and Virginia Streets reflect Glasgow's trade in sugar and tobacco with these two colonies.

This depiction was prepared at the peak of the tobacco

Charles Ross, *A Plan of Glasgow* (1773)

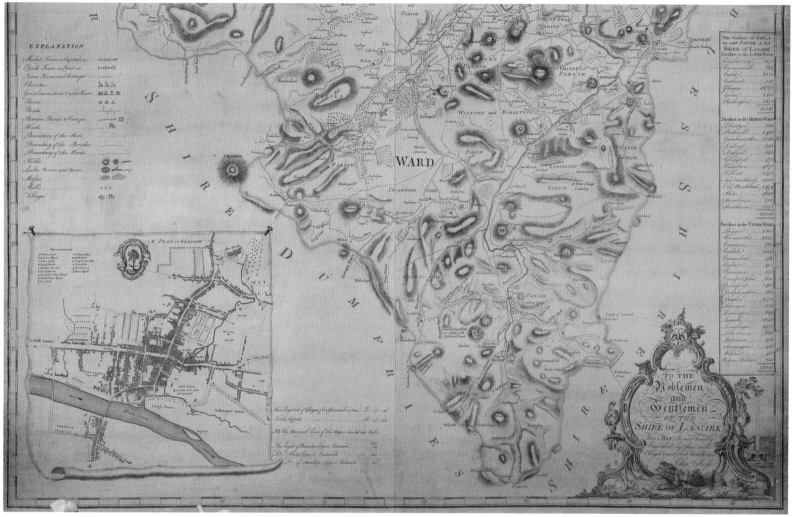

The unusual feature of the plan being 'pinned' to the Ross county map

trade and before the American Revolutionary War, when the city was responsible for all but two per cent of Scotland's import of this commodity. In 1774, more than 32 million pounds were landed at Greenock and Port Glasgow, largely for transhipment to Europe. Elsewhere, trade had a further impact on the city's layout. The improvements in the Clyde navigation, which included the new Jamaica Street Bridge begun in 1768, resulted in the disappearance of the horse ford, and the need for easier access caused the removal of both the

West and Water Ports. Land around the intended St James's Square is marked as 'laid out for building' and Gorbals is shown as a village of some size spreading southwards and along the routes to Paisley and Rutherglen. The Macfarlane Observatory, established in 1757 by the university on its Dovehill lands and equipped with instruments gifted by a merchant in Jamaica, is marked – a clear indication of the importance of astronomy to navigation and trade.

Charles Ross first issued his proposals to publish this

Crest of the city arms above the town plan

Title dedication of the county map

county map in 1771, with a scheme relying on guaranteed subscriptions to support the undertaking. His original intention was to have the map ready for publication in the October of the following year at a cost of 12/-. The common practice of securing patronage from subscribers had an influence on what is shown on the resulting maps. For example, only three of the 21 estates owned by leading Glasgow merchants in the immediate area of the city are missing from the depiction of the county. However, this method was prone to the vagaries of the financial situation of the day. Several proposed mapping projects foundered on a lack of sufficient support from the local gentry and the delay in the eventual publication of this map may have been related to problems in raising funds. Certainly, in June 1772, the collapse of the Ayr banking firm of Douglas, Heron and Company had a serious impact on the finances of several Glasgow merchants.

The newspaper advertisement makes no reference to the map containing a plan of the city but, more importantly, the four sheets in which it was originally published were the work of George Cameron, an Edinburgh engraver, who at this date resided in Borthwick's Close in the capital. It may have been the case that Glasgow still did not have a press large enough to print such a map. Subsequently, Cameron was to engrave Ross's county map of Dunbartonshire. In 1775, Alexander Baillie republished the Lanarkshire map as a one-sheet reduction but without the plan. This date may mark the start of his time working in Glasgow with James Lumsden.

Ross remains a somewhat intriguing figure in the history of Scottish map-making. Although his cartographic style appears naive, suggesting that he may have been self-taught, it is possible that family and masonic connections enabled him to pursue a long and, presumably, successful career which continued into the nineteenth century. This was one of four county maps he produced and he was employed on numerous estate surveys, particularly in the west of Scotland. He established a nursery business at Greenlaw, on the outskirts of Paisley, published a *Traveller's Guide to Lochlomond* in 1792 and was something of an amateur archaeologist.

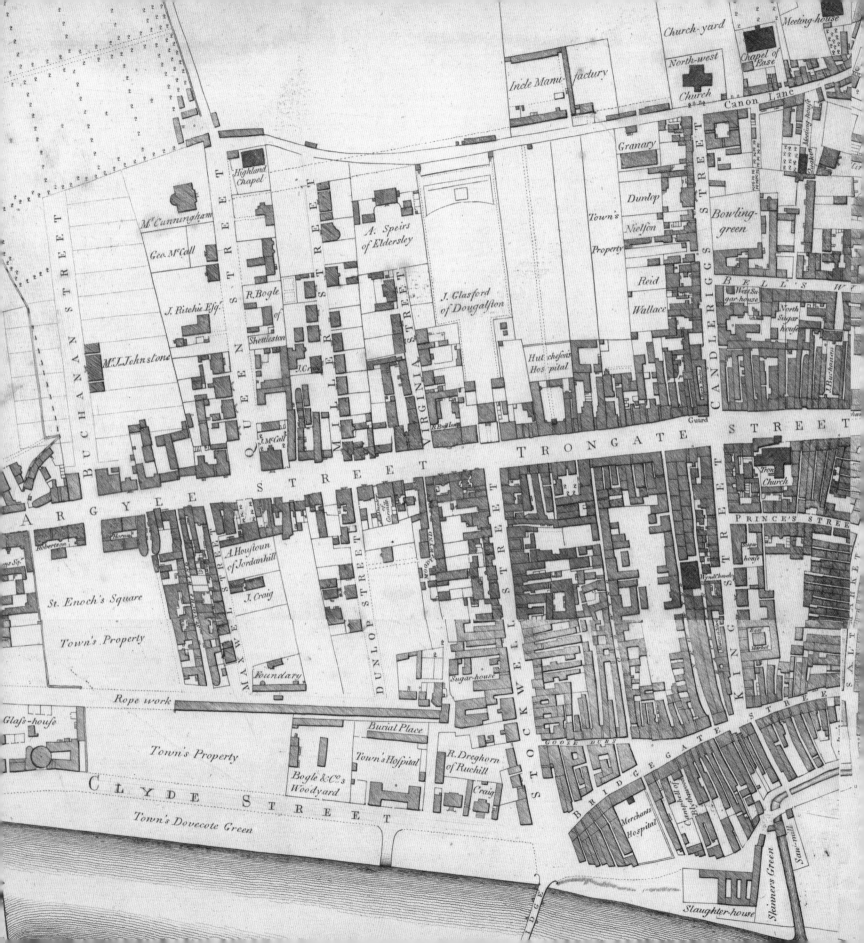

1778

'With all other Remarks proper to be set down': McArthur's plan

1778 is a comparatively late date for this, Glasgow's first comprehensive plan, but it should not be seen as an indication of a slower development in surveying within the city. As early as 1727, the civic authorities had employed John Watt on several surveys, which included a map of the lands of Provan and the Coshneoch Muir, while James Barry succeeded him in a varied range of commissions in and around the city. It was, however, John McArthur who prepared the earliest surviving map to depict the whole structure of the burgh and its component parts in greater detail.

This survey is a fascinating amalgamation of the old and the new, reflecting the dynamic changes happening in Glasgow. While some urban historians have commented on the contrast between the medieval pattern of the rigs and backlands of the old town and the more regular grid of the new streets developing off Trongate-Argyle Street, McArthur does not give a very clear pattern of this later geometric layout. What he does provide is an image at a sufficiently large scale to display many important elements of this development and growth.

Overall, the impression is still of a small town surrounded by gardens, parkland and fields but the identification of many of the burgh's significant buildings emphasises transition. Within the older city core, features of Glasgow's role as a local commercial centre survive, as signified by the herb, mutton and beef markets, the mills on the Molendinar Burn and the tanworks. Traditional industries such as a coachworks and brewery are marked, although the latter is located off St Mungo's Lane. It is worth noting that McArthur does not identify the more celebrated Drygate brewery established by Hugh and Robert Tennent. Of greater significance is the location of four sugar houses testifying to the importance of this commodity to the city's early prosperity. While tobacco was largely re-exported, sugar was processed in Glasgow itself, with rum distillation as a subsidiary activity.

John McArthur, *Plan of the City of Glasgow: Gorbells and Caltoun* (1778)

Newer industries can be discerned in the mapping of bleachfields, washing greens, an inkle factory and cudbear works (inkles being linen tapes used to trim clothing, and cudbear a purplish-red dye obtained from certain lichens) – an indication that, by the 1770s, Glasgow was Britain's leading linen town. Elsewhere, tile works, a foundry and pottery suggest the gradual industrialisation of business. In addition, key features of the future shape of the city can be identified in the location of certain premises. While several enterprises are opening up on the city outskirts, others continue to remain in the heart of the burgh. For example, the land to the west of Jamaica Street is occupied by the muslin manufacturing business of Brown, Carrick & Company while the north bank of the Clyde already has a glass factory, wood yard and ropeworks separating the desirable residences off Argyle Street from the river.

Among the leading citizens named in the 'new town' are John Glassford, Andrew Houston, George Bogle and Alexander Spiers, while the Duke of Montrose's Lodgings on Drygate and Dr John Moore's lands beside the College grounds suggest that there had not yet been a complete flight away from the centre by genteel society. A solitary crane on the Broomielaw would seem to be the only sign of any wharf or harbour facilities. Although St Andrew's Church still stands isolated from any other buildings, McArthur does show

nascent development in the east of the city, particularly the ground laid out on John Orr of Barrowfield's lands leading directly from the newly built Rutherglen Bridge.

Unlike many of its contemporaries, the decoration on this plan is not overly ornate, with a relatively simple municipal crest and dedication to the civic authorities. However, McArthur continues the subtle emphasis of contrast. He alludes to the city's history by ornamenting the title with a sketch which includes the Cathedral, the ruins of the Bishop's Castle and the Dean's manse. Offshore, images of a small coasting vessel and a three-master recognise the source of the city's growing prosperity. The picture is completed by an apparently random collection of casks, bales, packages and agricultural implements strewn in the foreground.

Surprisingly, the first record of McArthur's proposals to publish this plan appeared in October 1778, by which date he must have completed and printed his survey since, in the same month, he sent a proof of the plan to the Council. This resulted in the purchase of ten copies and his nomination to the list of burgesses and guild brethren. Four years later, a further twelve copies were purchased, coloured to indicate parish boundaries, for the city's clergy. While he is possibly best remembered for his work surveying the south side of Loch Tay for the Earl of Breadalbane, McArthur was also employed by the Commissioners of the Annexed Estates and, more significantly, by certain leading Glasgow citizens, including John Campbell at Killermont in 1776 and James Dunlop at Heateths, the following year. Dunlop was the son of Colin Dunlop, provost of Glasgow between 1770 and 1772, while Campbell held the office between 1784 and 1786. McArthur's work for them may have made the Council amenable to his talents. Certainly, he had been resident in the city since 1768, supplementing his income by teaching mathematics and land surveying, as well as producing leather goods and renting out summer lodgings. Similarly, Alexander Baillie, the map's engraver, also offered the teaching of drawing. The following year, Baillie was to publish a single-sheet reduction of the plan, possibly as a cheaper, more portable and more marketable product.

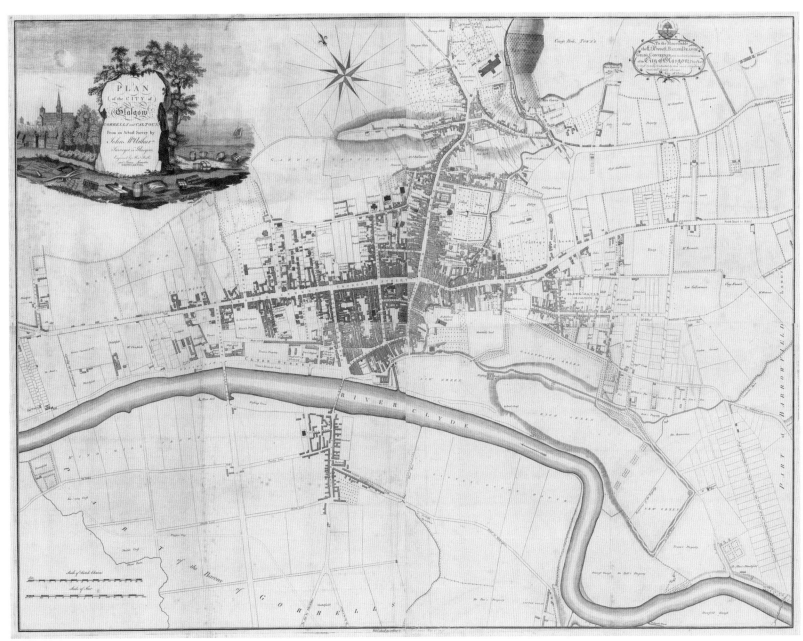

The complete map emphasises how little Glasgow had grown by this date

1782

A Council commission to delineate Glasgow's boundary

Throughout the mid-eighteenth century, one name appears regularly as the preferred surveyor, employed by both public institutions and private individuals, to record the changing layout of Glasgow. This was James Barry, who had moved to the city in 1734. By the early 1750s, he was working as collector to the Merchants' House and was responsible for a plan of its lands at Wester Craigs, before becoming involved with similar commissions at Petershill, the Easter Common, Carmyle and Summerston. As Glasgow began to expand outwards, he laid out many of the new streets – Argyle, Jamaica, Virginia, Miller and Dunlop, as well as Gallowgate in the east. He also supervised the construction of the Jamaica Street Bridge across the Clyde. In fact, he has been credited with creating the markedly geometric grid plan of what is now known as the Merchant City. This grid was, of course, no new thing but sprang naturally from the field and rig boundaries imposed on a landscape with little in the way of dramatic features.

His career is a valuable example of a city requiring, commissioning and using plans for a range of administrative purposes. He first appears in the Council records in 1759 and, fourteen years later, was confirmed as 'surveyor and measurer for the city', reflecting a confidence in his abilities across a wide range of duties, which included levelling, lining and repair. The details of his appointment specifically mention 'preventing encroachments'. Four years earlier, he had participated in an inspection to determine the city's eastern boundary with the Barrowfield estate which required a correction westward. Urban growth had an impact on land values and Barry fulfilled the role of trusted expert.

This depiction of the city springs directly from the Council's requirement to ensure that its boundary was clearly defined. Early in 1776, a plan was authorised to be purchased from Barry 'so as the extent of the royalty of this city may be knoun and preserved'. Its importance may be underlined by

James Barry, *Plan of the City of Glasgow, Gorbells, Caltoun and Environs* (1782)

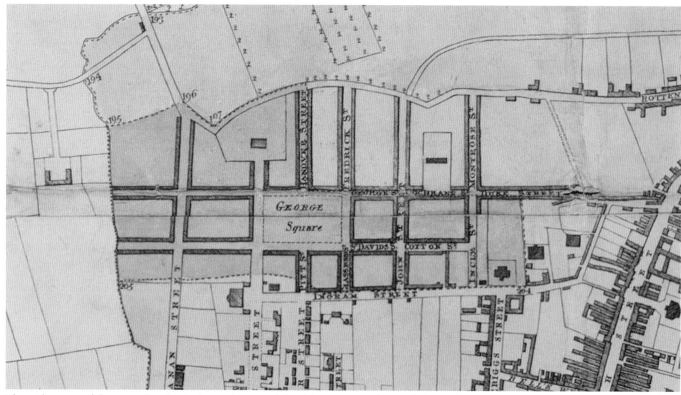

The grid pattern of the town plan developing to the west of the old town, centred around George Square

the Council undertaking responsibility for the engraving. In 1777, John Gibson published his *History of Glasgow, from the earliest accounts to the present time* and included a rather general plan of the city as a frontispiece, specifically engraved by Andrew Ready from this survey, suggesting its completion by that date. However, following a perambulation of the royalty in 1780, amendments to the positioning of certain stones, which indicated the boundary line, held back the final publication for another two years.

Once again, a map user needs to be wary in how the information displayed is interpreted. What looks like the record of new streets opening up in different sectors of the city is more a sign of intent. It is important to remember that much of the surveying on the plan could date from at least five years earlier and that this was a commission prepared for a specific admin-

istrative purpose by someone in the Council's employ. Because it is a demarcation of the boundary, it covers a much wider area than other contemporary plans, most notably that by McArthur, and is drawn at a smaller scale, thereby affecting the detail included.

By this date, the Cathedral, which had held power for much of Glasgow's early history, now appears isolated from the main hub of the city. More significantly, Barry has taken pains to identify his own handiwork in providing a far clearer impression of the plan developing to the west of the old town. This is particularly true in his scheme for the lands of Ramshorn and Meadowflats, Glasgow's first example of a deliberately planned development as a civic exercise. Earlier street layouts had been characterised by somewhat tentative, piecemeal promotions financed by individual citizens and prone to limitations but this

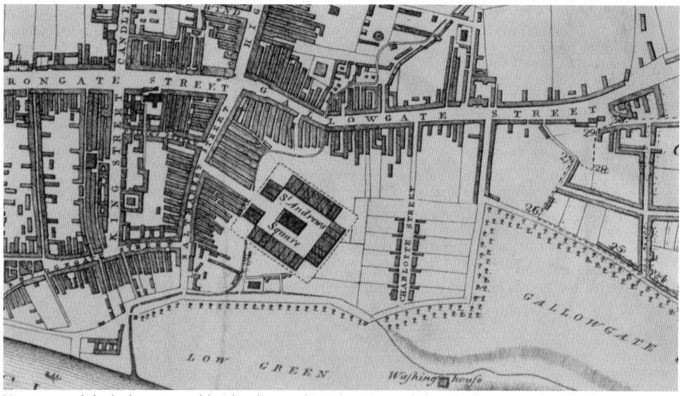

Nascent proposals for development east of the Saltmarket around St Andrew's Square which pre-dated construction by at least four years

depiction clearly identifies a bold, regular plan of solid frontages centred on George Square and around St Andrew's Church. None of this grid was a reality when the map first appeared. In St Andrew's Square, building construction did not begin until 1786, when Barry was responsible for marking out the individual parcels of land to conform to William Hamilton's plan. Furthermore, while the Council had bought the Ramshorn lands from the Hutchesons' Hospital patrons in 1772, the collapse of the Ayr Bank on the day following debate on the purchase price dramatically affected the sale of plots, which dropped from 4/- to 2/6d per square yard within a month. Development stalled; for example, Montrose Street, the eastern edge of this design, was not opened until 1787 and the proposals for a new suburb based on St James's Square foundered completely. Regardless of this, the Council clearly trusted Barry's

work, for a copy of the plan was still in use in a subsequent boundary inspection forty years later.

Given Barry's involvement with transport developments in Glasgow's hinterland, it is significant that he is the first to identify a number of ships on the Clyde, concentrated along the Broomielaw. He also indicates the Port Dundas branch of the Forth and Clyde Canal, built to secure the financial support of Glasgow merchants who feared losing business if the canal bypassed them completely. However, this terminus was not established before 1786. It is also interesting to note that, while the title of the plan bears a close similarity to McArthur's work and both were engraved by Alexander Baillie, there is no obvious reliance by either surveyor on the other's depiction. This appears to be Baillie's last piece of work in Glasgow before he returned to the capital, where he died in 1791.

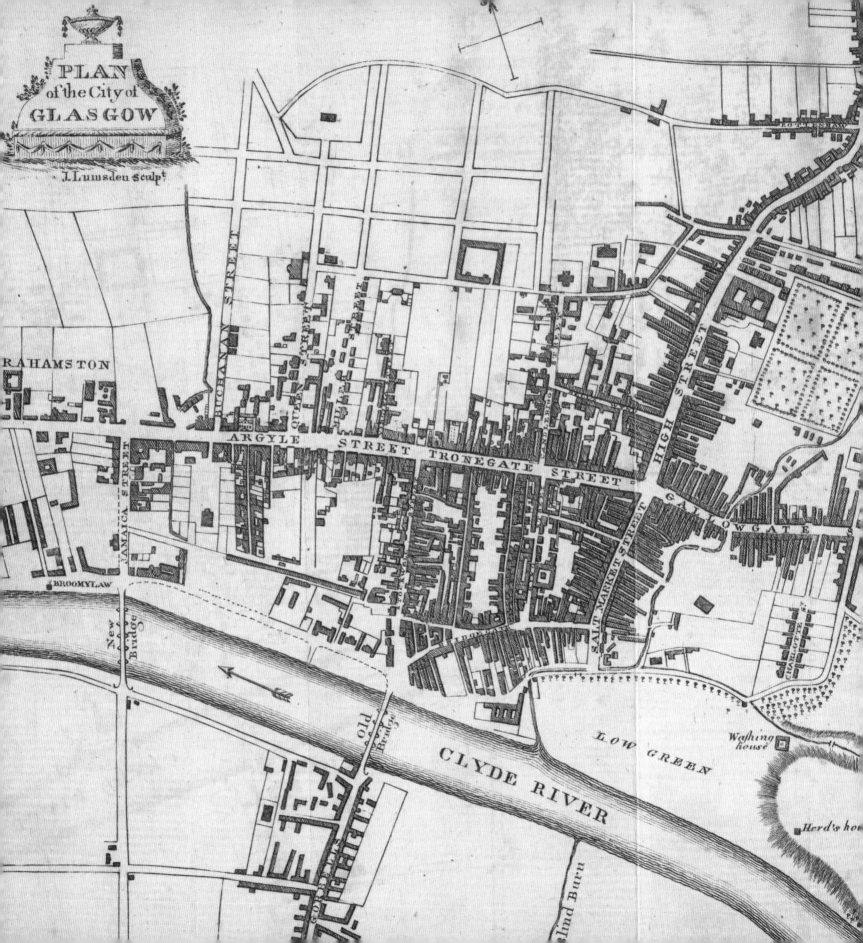

PLAN
of the City of
GLASGOW

J. Lumsden sculpt

RAHAMSTON

BUCHANAN STREET

QUEEN STREET

JAMAICA STREET

ARGYLE STREET TRONEGATE STREET

HIGH STREET

GALLOWGATE

SALT MARKET STREET

BROOMYLAW

New Bridge

Old Bridge

CHARLOTTE

LOW GREEN

Washing house

CLYDE RIVER

Herd's hou

Blind Burn

1783 & 1790

Two early 'directory' maps – two different visions

Glasgow's growth in population and prosperity led to the development of a more sophisticated consumer market. Booksellers and publishers began to provide an increasing variety of literature to its citizens, with the second half of the eighteenth century witnessing a dramatic rise in the number of books and pamphlets published in the city itself. In December 1783, John Mennons advertised his *Glasgow Almanack* for 1784, embellished with 'an elegant plan of the City ... engraved on purpose for the work'. Mennons (1747–1818) had originally been apprenticed to an Edinburgh printer and had set up in business for himself, in 1777, in the Lawnmarket, producing a variety of journals which offered

James Lumsden, *Plan of the City of Glasgow* (1783)

readers both information and entertainment. In 1782 he moved to Glasgow and, on 27th January 1783, published the first issue of the *Glasgow Advertiser* from Gibson's Wynd in the Saltmarket. His intention was for it to be free from political prejudice and it is noteworthy that the *Advertiser*, forerunner of the *Glasgow Herald,* retained its independence and impartiality. The accompanying plan, engraved by Lumsden, also appears in Mennons's *Glasgow Magazine and Review; or Universal Miscellany* of 1783. While it is recorded that James Lumsden founded a printing and engraving business in the city in that year, there are several earlier examples of his work, including Ross's map of Stirlingshire of 1780 and William Semple's 1781 plan of Paisley. His son was to be responsible for publishing several Glasgow plans in the first half of the following century. Although a version of James Barry's survey was used to illustrate Gibson's history of the city in 1777, this is the first plan engraved specifically to accompany a publication designed for Glasgow inhabitants.

Later commentators have asserted that the 1783 plan is a revised and updated version of McArthur's map, with the first indications of the New Town proposals becoming clear. A careful inspection of the details would suggest a similarity of style which more closely matches Barry's 1782 plan. This is, however, no copy, for significant differences exist – most notably in the pattern and alignment of the Ramshorn grid, where there is no clear identification of George Square and certainly no block drawing of the building frontages. Additionally, the St Andrew's Square development has been removed completely, leaving Dreghorn's church isolated from all other buildings. A smaller scale and size has resulted in a certain loss of detail and, while the plan would have been a conveniently sized, portable depiction of the city's layout, ideal for any visitor, it could also be a reflection of a less assured approach to the engraving. While some might discern a sign of a decline in business confidence, as exemplified by the lack of any vessels on the Clyde, it is equally possible that the simpler cartographic style represents Lumsden's own abilities. His partner, Alexander Baillie, had returned to Edinburgh in 1782. The plan is, in fact, Barry without the proposals and, as it was prepared at a time when the city was recovering from both the financial crisis of the 1770s and the impact on its trade of the American Revolutionary War, it perhaps reflects a more sober or pragmatic vision of the city.

Mennons was also involved in the publication of *Jones's Directory, or, Useful Pocket Companion* which listed the city's merchants, manufacturers, traders and shopkeepers. Although it ran only from 1787 until 1793, a plan of the city dated 1790 'from actual survey' was included in the volume for 1791. Partly because it has rarely been discussed, this is a particularly interesting view of the city as the eighteenth century was drawing to its close. While there is no attribution of either surveyor or engraver, the title cartouche is similar to that engraved in 1783 but, possibly in recognition of a growing prosperity through trade, two vessels now adorn the illustration. Prepared at a noticeably larger scale and size, and covering a wider hinterland, this plan may reflect a growth in self-assurance over the intervening years. The contrasts with its predecessor are certainly noticeable. More individual buildings are identified; for example, the Grammar School, later the High School of Glasgow, in George Street, the Anderston brewery, the Theatre Royal, opened in January 1782 in Dunlop Street and the Thistle Bank, founded in 1761 and one of six banks listed in the directory, in Virginia Street. A bowling green still exists on Candleriggs but, uniquely, this plan marks the tolls on all the major routes into the city. The route to the canal basin is once more identified and may reflect the completion of the Forth and Clyde Canal in 1790. Ships are again berthed at the Broomielaw.

Overall, the depiction returns to a closer reliance on the Barry plan and hints that the city was entering a period of recovery and growth. The Ramshorn–Meadowflats scheme replicates his original layout and, while not an exact match,

OPPOSITE. An entirely different layout in the St Andrew's Square area, as well as shipping at the Broomielaw, is shown on this plan

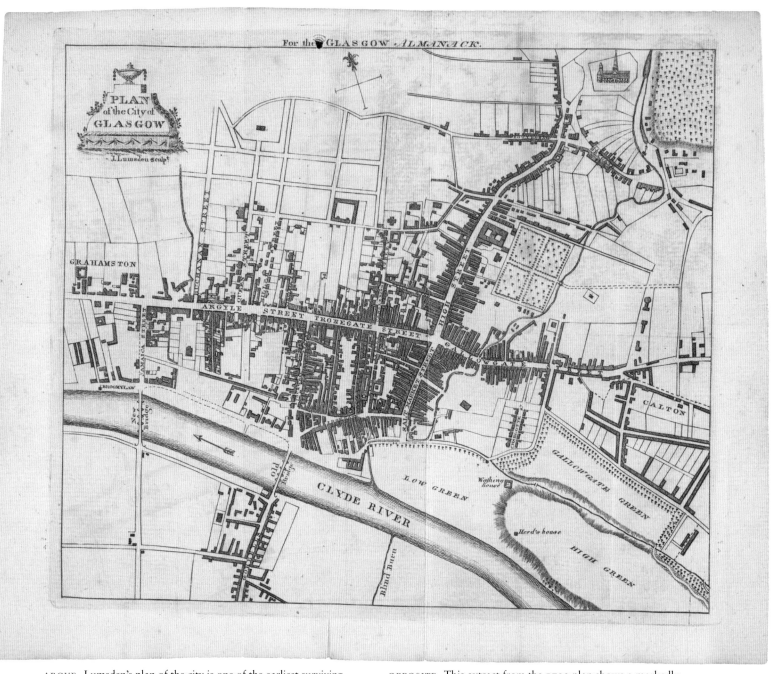

Plan of the City of Glasgow

For the GLASGOW ALMANACK.

PLAN
of the City of
GLASGOW

J. Lumsden sculp.

GRAHAMSTON

ARGYLE STREET TRONEGATE STREET

HIGH STREET

GALLOWGATE

BROOMYLAW

New
Jamaica
Bridge

Old
Bridge

CLYDE RIVER

Blind Burn

LOW GREEN

Washing house

Herd's house

HIGH GREEN

GALLOWGATE GREEN

CALTON

ABOVE. Lumsden's plan of the city is one of the earliest surviving
directory maps

OPPOSITE. This extract from the 1790 plan shows a markedly
different layout around George Square

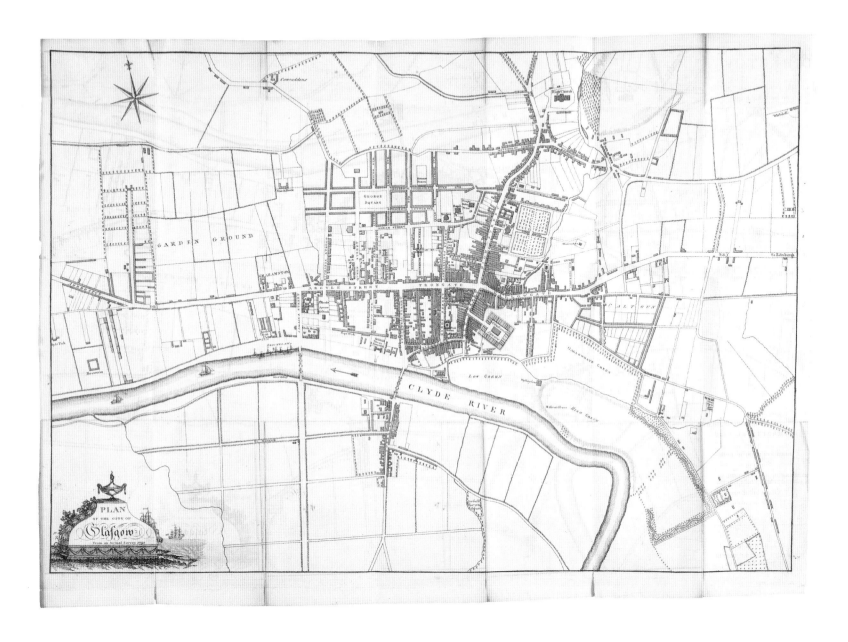

the buildings around St Andrew's Square are again shown in plots. There are no recorded plans of the city between these two depictions which, in itself, may suggest somewhat less dynamic activity in the period. Building speculation suffered from a severe shortage of capital and potential investors, combined with a complicated feu contract process. Not unsur-

prisingly, neither plan replicates Barry's indication of the stones set down to indicate the burgh royalty but, more significantly, nor do they include the city crest or a dedication to the civic authorities. The 1791 directory contains the last record of Barry, still listed as a land surveyor, residing in Carsbasket's Land, Gallowgate. He died in the following year.

	N°			N°			N°			N°
...ing	53,	Cuilhill		60,	Blackhouse		67,	Patonswells		
...nonhead	54,	Bruntmuir		61,	Bentgate		68,	Summerlees		
...y Know	55,	Garnheights firs		62,	Green		69,	Williams Taylors		
...nloning	56,	Drumpellier farm yard		63,	Langlone		70,	Sandy Know		
...nshill	57,	Hot house		64,	Old Quarry		71,	Langlees Planting		
...adie	58,	Drumpellier mansion house		65,	Mr Langmuirs		72,	Thomas Adams		
...muick	59,	Blairhill firs		66,	Marystown		73,	James Smith		

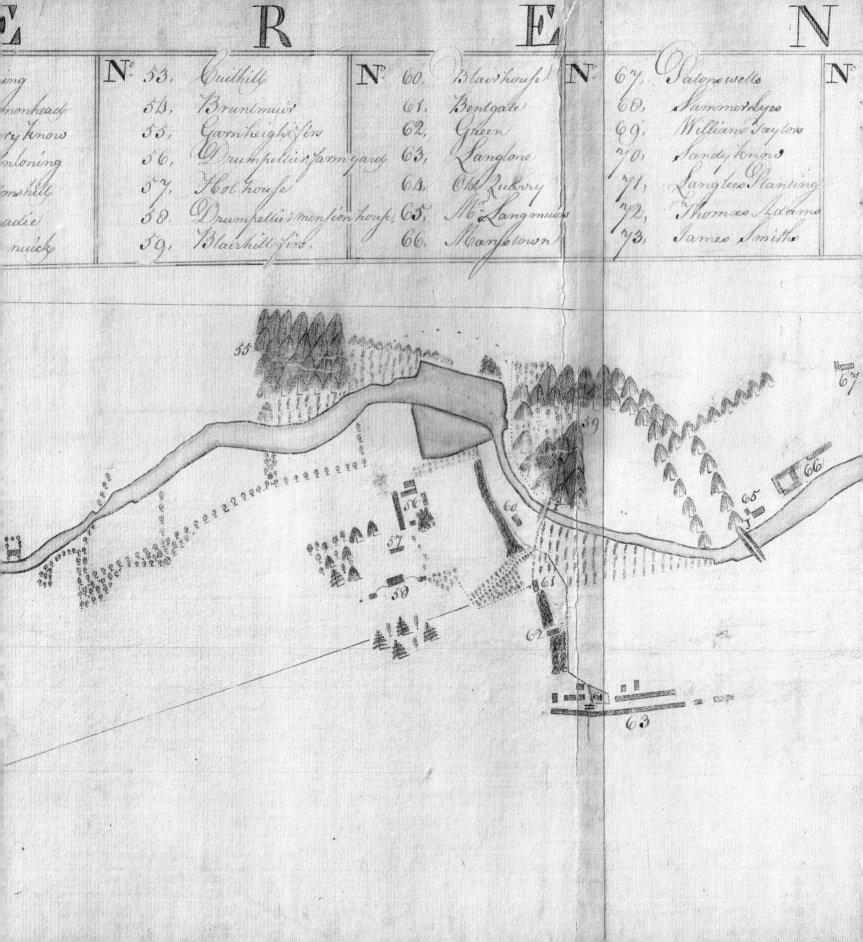

C.1790

A transport success story: the Monkland Canal

Although a project of relatively short distance (a little under 20 kilometres), the construction of the Monkland Canal proved to be a long and difficult enterprise. Glasgow was surrounded by areas with rich coal reserves and, as the city grew, the demand for this fuel increased. Faced with steadily rising prices, the Council, Trades House and Merchants' House had all considered the issue, coming to the decision that the most effective answer was to create a canal link to the coal-fields. Under the influence of Provost James Buchanan of Drumpellier, it was probably inevitable that such a link would be to those reserves in the Monklands area.

James Watt, who had been employed already on surveys relating to improvements in the Clyde navigation, was commissioned in 1769 to report on a suitable route and, of his two proposals, the Council opted for the cheaper alternative without locks, terminating just outside the town and using a wagon-way with a self-acting incline to connect with Glasgow itself. A subscription list was opened and powers to begin construction were obtained by a local act of parliament in April 1770. Watt was appointed Resident Engineer and, between May 1770 and July 1773, supervised the work, beginning at Sheepford in Old Monkland and working west. This was to be his only major engineering commission and errors in both his levelling and assessment of the soil conditions suggest that he was not ideally suited to the role of a practical surveyor. Heavy clay soil, poor workmanship, inclement weather and financial difficulties all plagued the progress of the work, which was done in sections. When completed, these were used to transport construction material or coal. Additional building costs and poor book-keeping appear to have swallowed up much of the subscription money invested in the project and, while it may not have been the cause of the cessation of work, the collapse of the Ayr bank in 1772 did nothing to improve the general financial situation.

Hugh Pate, *A Map of the Monkland Canal* (c.1790)

43

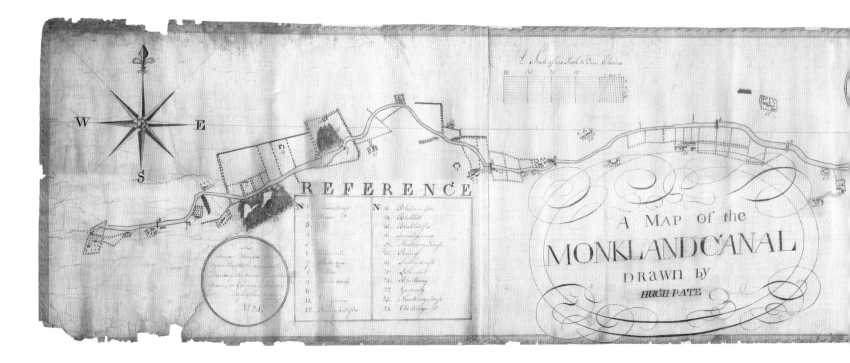

With only a little more than seven miles of the canal completed, Watt ended his connection with operations the following year and moved to Birmingham in May 1774.

Unfortunately, revenue on the unfinished canal failed to meet the costs of its upkeep. In addition, several of the original promoters had been badly affected financially by the American Revolutionary War. In consequence, the canal was sold at public auction in 1781 and the new owners resolved to extend the canal to a point nearer the city at Townhead, which involved the building of a series of locks at Blackhill. Subsequently, the canal's existence was assured by an agreement to supply water to the Forth and Clyde Navigation. An eastward extension to the heart of the coal-field at Calderbank was built and a cut of junction was constructed at Hamiltonhill to link with the Forth and Clyde Canal at Port Dundas. Work at the western end was completed in October 1791 and in the east by 1794. While shareholders saw no return until 1807, the canal was a financial success. Revenue steadily increased from this date, particularly with the development of iron and steel manufacture in the Coatbridge area. Although passenger traffic was never of primary importance on it, over 30,000 people were being carried each year by the 1830s, and by 1850 the annual transport of coal was in excess of a million tonnes. Competition from neighbouring railway lines, particularly after the late 1860s, brought an inevitable decline in traffic and in 1942 authorisation was sought to abandon it. The canal became seen as an obstruction to development, a danger to children and an eyesore. In the 1960s, most of the route was filled in and, subsequently, the section from Townhead to Easterhouse was buried under the M8 motorway. Water from it still feeds the Forth and Clyde Canal through culverts.

While this plan is undated, it was the work of Hugh Pate and must have been prepared in the period after the extension of the canal west from Gartcraig since it shows in detail the line to the 'Bason House'. Conversely, it does not show anything of the cut of junction joining the Monkland and Forth and Clyde canals. In addition, there is no indication of the locks which were constructed at Blackhill. This would place the preparation of the plan somewhere between January 1787 and October 1791. Nonetheless, this is a most valuable

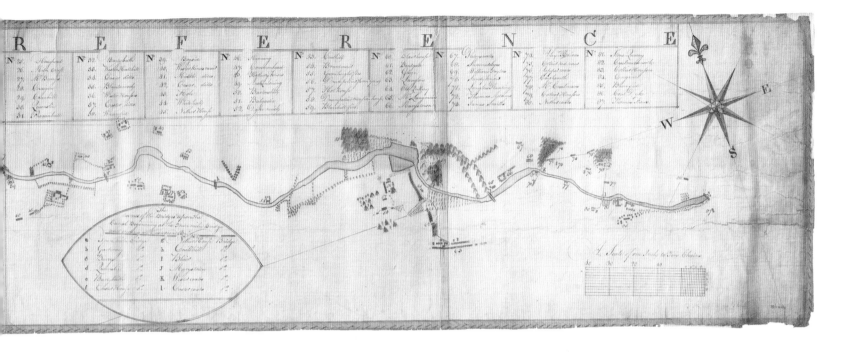

document in its provision of information on the bridges and neighbouring properties along the route of the waterway. A good example of this is the identification of the reservoir at Drumpellier, created to provide water for the canal. Interestingly, the plan shows little of the coal works which were the reason for its construction. Other than the location of colliers' houses and an engine, only two quarries are marked and the greater focus appears to be on properties and plantings. Comparatively little is known of Pate or his surveying work. His only other surviving survey is a plan of Palace Craigs and Monklands, dated April 1790, suggesting that he was employed in the immediate area at this time. It appears likely that he was a local man as a record of his baptism is listed in the Old Monkland register for June 1768.

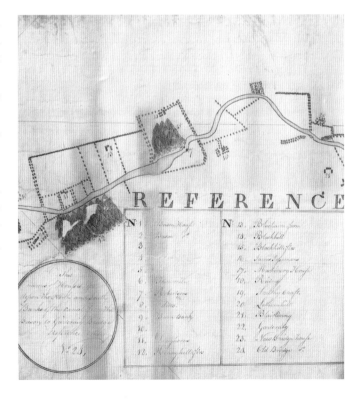

ABOVE. The whole map focuses on showing properties adjacent to the Canal

RIGHT. Fir plantations at Kennyhill and Blackhill are part of the map's detail

Eastfield

Cockmoor

Balornock

Port

Boundary

Craufurd Esqr

Springvale

of Royalty

Balgray

Fairhill

Kepoch

Balornock

Barnhill

Peterſhill

Jaſ Campbell Esqr

Broomfield

Boundary of Royalty

Cowlairs
Scott Esqr

Jnᵒ Hamilton Esqr

Kelvinſide
Dr Lothian

N. Woodſide
Rⁿ Barton Esqr

Hamiltonhill

Germiſton
Dinwiddy Esqr

Mill

Kelvin Water

Roſebank
Armour Esqr

Hillhead

Southwoodſide

Craighall

Pinkstonbog

B

Sawmillfield

Hundredacrehill

A

R

O

N

Blochern

Kelvingrove
Patison Esqr

Port Dundaſs

Broomhill

Gairbridge

Blocha

Jⁿ Willson Eſ

Cowcaddens

Portfield

Linehouſe

Monkland Canal Esqr

Sandyford
Wm Frensh Esqr

Glasgow field
Thoſ Steuart

St Rollocks
A. Warren

Overnewton
Rt Craig Esqr

Saughyhall

St Enochs Burn

Y

Town Mills

Millbank

Chr

Finnieſton

Napershall

G

I

Kirk

Town Mills

Craigs

Ko

Stobcroſs
Phillips Esqr

Anderſton

Quarry

Cudbar
McGrouh

Hyde Park

Whitehill

W. Gordon Esqr Haghills

Langfield

McAuhorn Eſ

Annfield

Gateſide
Mr Donaldson

Mavisbank
Jaſ Hamilton Esqr

Trades Town

Sfield
Miller Esqr

head

Plantation
wen Esqr

Conynghouſe

Campblefie

Mair Esqr

Parkhouſe
Wm Croſs

Huchiſon Town

Jean field
Jaſ McehoseEsqr

Standingstone

Gorbals

Green

Barrowfield

Newlands
Jaſ Mcehose Esqr

Toll

Dockenfaulds

Graham Esqr

Springbank
Jaſ Wardrop Esqr

L. Shiels

Newhouse

A

R

I

S

H

Shiels

Moorhouses

Belvide
R McNair

Tiltwood

Butterbiggings

Flugsbowes

Little govan
Rt H. Rae Esqr

Dye Work

Weſthor
Jaſ Denniston

Flugs Castle
in Ruins

Strathbungo

Allanton

½ Miles

erfield

Croſsmyloof
lands

Thoſ Crawford Eſ

Govan Coal Work

Dalmarnock

Wardrop Esq
Dai

Battlehere 1555

Croſshill
Jⁿ Rowen Esqr

Polmadee

Shawfield

Jⁿ Buchannan Esqr

Rutherglen Farm

1795

'Nobody here entertains any doubt of the advantage of turnpike roads'

So wrote John Naismith, subsequently author of the *General View of the Agriculture of the County of Clydesdale*, in the *Statistical Account* of the parish of Hamilton in 1792. This map of the country seven miles round Glasgow shows the city in the context of its immediate hinterland and, while Thomas Richardson did not specialise in road surveying, it does give a clear impression of a growing road network. It was published at a time when the turnpike road system was beginning to develop more extensively in Scotland as a consequence of increasing trade and travel. While more effective legislation had been in place since the middle of the eighteenth century, real progress seems to have been delayed until its later decades. Between 1772 and 1798, there were twelve separate parliamentary acts relating to Glasgow roads alone and eventually there would be 22 turnpike trusts in Lanarkshire. The individual parish reports for the county in the 'Old' *Statistical Account* give an impression of a markedly variable road system throughout Lanarkshire, with the quality and condition of routes in more remote areas remaining poor. Understandably, roads nearer Glasgow itself were better maintained and it has been suggested that this was the single most important factor in the city's growth in this period. Not only are the tolls identified on the map but the key also states clearly that the distance from Glasgow Cross of each gentleman's seat is recorded, as well as the mileage of each village from the 'tolls next the City'. Three years later, Richardson was to include maps of the roads from Glasgow to Paisley and Dumbarton in his *Guide to Loch Lomond, Loch Long, Loch Fine*.

The scale of the map is rather too general to allow a detailed depiction of features within the city but, significantly, it provides the first indication of the new suburb being laid out south of the Clyde in Trades Town, while Hutcheson Town is identified by name only. Some of what Richardson does show is open to question. Although the new town layout

Thomas Richardson, *Map of the Town of Glasgow & Country Seven Miles Round* (1795)

around George Square is mapped, the only city building named is the Cathedral and only as 'kirk'. Immediately to the north is marked a gentleman's seat and not the new Royal Infirmary opened in December 1794. In contrast and perhaps again emphasising the focus on routeways, Richardson does includes the Hutchesontown bridge which was commenced in 1794 but was washed away by floods in the following November while still being built.

Richardson's career provides an encapsulation of the experiences of many a surveyor of his day. He could be said to have started well by his being apprenticed to John Ainslie, the leading Edinburgh cartographer, whose work included surveys of several Scottish counties. Richardson was to gain a wide experience working under Ainslie between 1786 and 1791, at a time when the latter was focussing on his great map of Scotland and before commencing a detailed survey of the Eglinton estates in Ayrshire. In October 1792, only one year after finishing his tutelage, Richardson showed his mettle by announcing his intention to publish a map of the Glasgow area by subscription. While there is no surviving record of those subscribers, an alphabetical list of more than 140 country houses with their owner's names flanks either side of this map. As each property is shown and there is a notice of

a reduced price for subscribers, it can be assumed that the names are those who contributed to its cost. However, additional estate and owner names have also been recorded on the map. The map itself was also used to advertise the surveyor's services, with a note stating 'Estates accurately surveyed ... Plans copied, diminished or enlarg'd'.

A valuable insight into contemporary attitudes to another aspect of the surveyor's work is provided by Richardson's newspaper intimation. While mathematical instrument-making in Glasgow can be traced back to the early decades of the eighteenth century, when craftsmen such as Henry Drew and George Jardin are recorded as repairing equipment used in lectures at the University, it is only with the establishment of James Watt's business in 1757 that the city could be said to have its first outlet for the surveyor's 'tools of the trade'. Despite this, many professionals preferred to purchase from London sources. Richardson himself specified particularly that he had 'procured from the first mathematical instrument-makers in London, the very best instruments for surveying and planning of grounds' when he first announced his proposal. Additionally, his preference for the Edinburgh engraver James Kirkwood was a careful choice to emphasise elegance and quality of workmanship.

Not every scheme met with success and Richardson's 1802 plan for a county map of Lanarkshire, also by subscription, got no further than the press notice. Furthermore, his dedication of this present map to Andrew Houston of Jordanhill could not have been more inopportune, since 1795 was the very year that the Houston family's West Indian sugar company collapsed, one of Glasgow's biggest business failures of the eighteenth century. Surveying itself could rarely support an individual in a career of any length. In Richardson's case, he was also proprietor of a map and stationery shop in Argyle Street from 1801 onwards, as well as being surveyor for the city's Trades House between 1819 and his death, ten years later, in 1829.

LEFT. The map key states the distances of properties and villages from Glasgow Cross

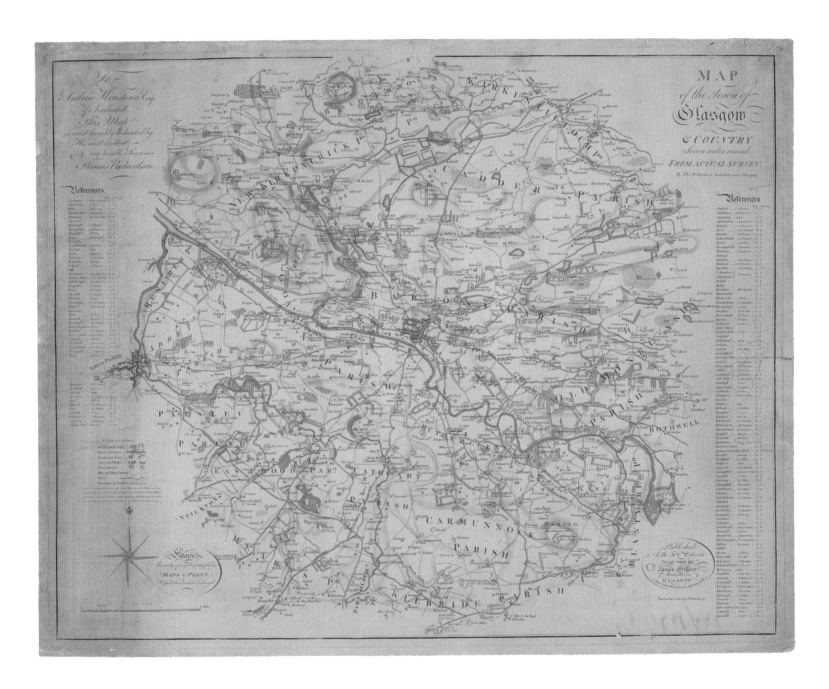

MAP
of the Town of
Glasgow
& COUNTRY
Seven miles round
FROM ACTUAL SURVEY
By Tho: Richardson Land surveyor Glasgow

To
Andrew Houstoun Esq.r
of Jordanhill
This Map
is most humbly dedicated by
His most obedient
& very humble Servant
Thomas Richardson

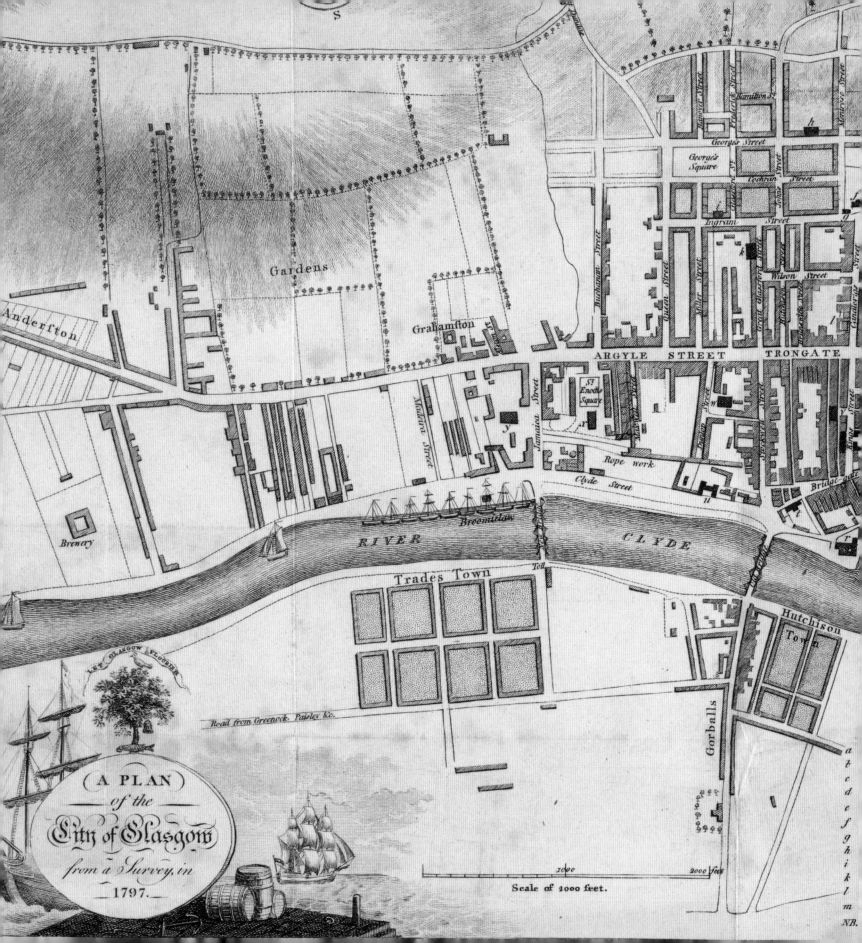

A PLAN of the City of Glasgow from a Survey, in 1797.

Anderston

Gardens

Grahamston

George's Square

George's Street

Hamilton Street

Frederick Street

Montrose Street

Cochran Street

Ingram Street

John Street

Buchanan Street

Wilson Street

ARGYLE STREET

TRONGATE

Jamaica Street

Madeira Street

Queen Street

Miller Street

Virginia Street

St Enoch Square

Rope work

Bridge-gate

Brewery

River Clyde

Breomielaw

Clyde Street

Trades Town

Toll

Hutchison Town

Gorbals

Road from Greenock, Paisley &c.

LET GLASGOW FLOURISH

Scale of 1000 feet.

1000   2000 feet

a b c d e f g h i k l m

NB.

1797

'Engraved in a very superior style from correct drawings'

James Denholm (1772–1818) was a man of marked abilities in several fields. A skilled miniaturist and landscape painter, by 1801 he was resident in McAusland's Land, Trongate. While he is said to have taught at the Glasgow Drawing and Painting Academy, located on Argyle Street, in the following year he was able to advertise an established curriculum of geography and associated subjects at his own Academy, 'first land east of Hutcheson Street'. To these, he was later to add 'the principles of perspective, the drawing of machinery, land surveying and the protracting of maps and plans'. Subsequently, he joined the Glasgow Philosophical Society and served as its president between 1811 and 1814.

His name first appears in 1796, when he intimated his proposal to publish, by subscription, a history of the parish and town of Lanark, describing himself as a writer of that burgh. A series of engravings from his sketches of some local country houses date from this year and may have been intended to illustrate the work. Although only 200 subscribers were required to provide a total capital sum of £50 and subscription papers were to be available in Glasgow, Edinburgh, Perth and Lanark, it appears to be yet another example of a scheme which failed. A similar plan to publish a history of the Clyde Valley in the following year also met with no success. These suggest that Denholm may have been originally from the Lanark area. Certainly, he had a working relationship with Robert Scott (1771–1841), who, although based in Edinburgh, had been born in Lanark himself and engraved at least one of Denholm's local illustrations. In addition, both men worked on engravings for the 1802 Glasgow edition of the poems of Robert Burns

This map appeared in Denholm's *Historical account and topographical description of the city of Glasgow and suburbs*, which was published first in 1797. As such, it was part of a growing trend in producing material which catered for the

James Denholm, *A Plan of the City of Glasgow* (1797)

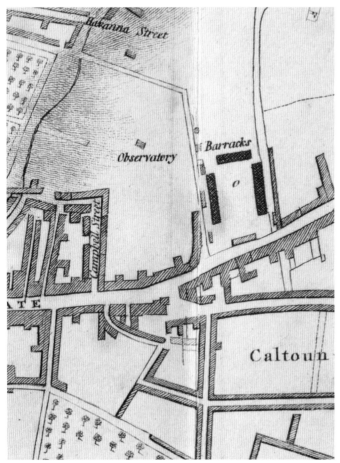

ABOVE. The newly constructed barracks built off the Gallowgate
OPPOSITE. The intended grid pattern of the new development in
Tradeston and Hutchesontown is shown for the first time on this
Denholm plan

needs of a developing tourist and travel market. The book was
printed by Robert Chapman, Stewart & Meikle and this is
recorded under the map, emphasising the press notice that 'to
render this work still more valuable, it will be ornamented
with a new and accurate map... engraved in a very superior
style from correct drawings taken on purpose for this publication'. Scott, described as the best Scottish engraver of his
time, was responsible for the production of the book's illustrations as well as the map plate. It was to be one of a series

of Glasgow plans he prepared, which included those in several
guidebooks, Peter Fleming's six-sheet map (1807) and David
Smith and James Collie's 1839 depiction of the burgh.

This is the first city plan to indicate the development of
suburbs south of the River Clyde in any detail. Growth was
directed both east and west of Gorbals village in a series of
regular projects on land owned by the Council, the Trades
House and Hutchesons' Hospital. Based again on the grid of
old field boundaries, both Trades Town and Hutcheson Town
are mapped as a series of regular rectangles with blocked
frontages and stippled central squares. While the text states
that some streets were already completed with two- to four-
storey houses, again this map suggests something more akin
to intent than reality. Elsewhere, many more buildings are
marked fronting the streets in the industrial villages of
Anderston and Calton. As the book was dedicated to the city's
Provost, James McDowall, it is not surprising that the burgh's
crest has returned to the title cartouche.

Overall, there is a similarity of style with the 1790
Lumsden map but Scott's talents can be seen in his clearer line
work and better definition. This is particularly true in the
shading he introduced to indicate the slopes of Fir Park and
Blythswood. Additionally, as the map accompanies a
guidebook, the engraver has marked the buildings in his table
of references in a darker shade to aid their easier identification.
In consequence, certain changes in society can be read from
what is shown. For example, the Duke Street prison, or 'New
Bridewell', whose first prisoners arrived in 1798, reflects
changes in the treatment of the wide range of offenders and
debtors, largely as a response to reforms suggested by John
Howard. As Denholm states, the introduction of solitary
confinement was seen as 'not only the most humane, but the
best calculated to answer one great end of punishment, the
amendment of the offender'. The barracks marked on the
Gallowgate were built in 1795. They played an important role
in accommodating troops at a time when threats of a general
uprising were underlined by riots in several parts of Scotland,
resulting largely from the Militia Act and the introduction of
conscription. In his text, Denholm comments that 'in respect

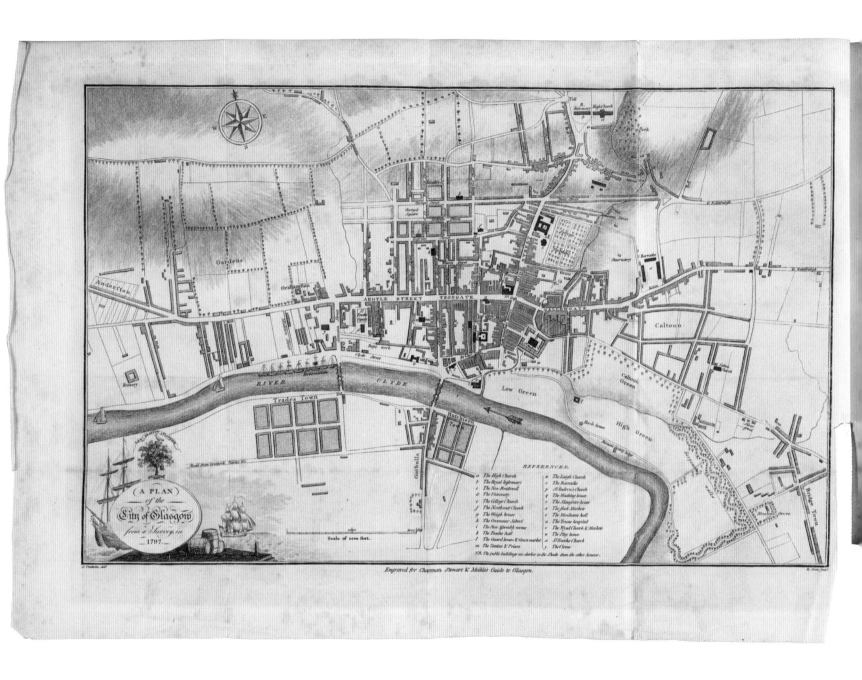

to the defence of the city, nothing is now to be apprehended from the designs of any enemies to the peace and good order of society'. Indicating other aspects of city life are the identification of the new Assembly Rooms, begun in 1796, and the Trades Hall, designed and built by Robert Adam between 1791 and 1794, both reflecting a growth in 'polite society'. At the other end of the spectrum, a circus is marked just off Jamaica Street.

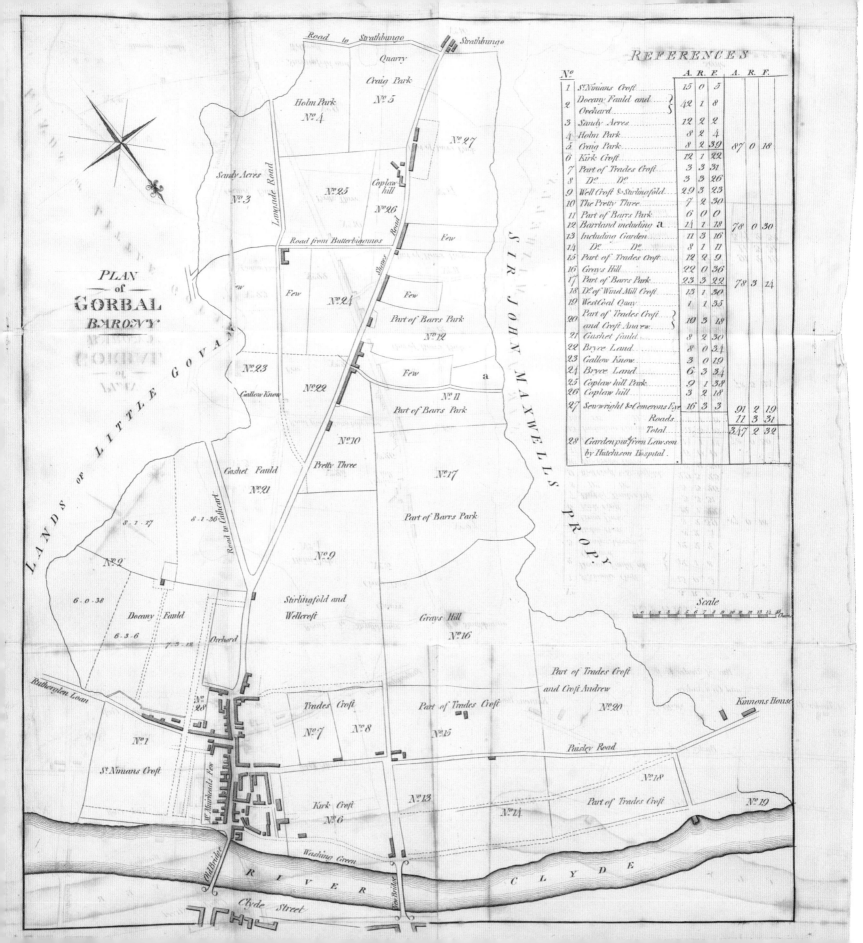

1803

Developing south of the Clyde

Despite an out-dated and not completely warranted reputation for poor housing and violence, partly a result of the novel *No Mean City*, the story of the Gorbals area, as part of the wider Govan parish, goes much further back. Situated at the south end of the lowest bridging point across the Clyde, it remained a small village outside the jurisdiction of the city for most of its early history. In 1650, ownership of the barony and regality was transferred, whereby the rents were paid to three beneficiaries, namely the Glasgow Council, the Trades House and Hutchesons' Hospital. Created a separate parish in 1771, by the last decade of the eighteenth century the population of the village, in line with the city's own growth, had almost doubled. Contemporary writers viewed the area as prime for development. The Rev. William Anderson commented in the *Statistical Account* that 'in 20 years a new Glasgow will probably be raised on the south side of the Clyde' and estimated that the population might grow to more than 50,000 inhabitants.

The initial plan to develop the lands resulted in a three-phase approach. The city magistrates took responsibility for Gorbals village itself and lands lying to the south, while the other two institutions established new suburbs in the eponymously named Hutchesontown and Tradeston. The Tradeston scheme of eight blocks was laid out, between 1790 and 1798, according to a plan by John Gardner, James Barry's successor as surveyor to the city, and was first to appear on Denholm's 1797 plan. Denholm himself commented that 'it certainly will be the finest village in Scotland' when completed. To the east, Charles Abercrombie was employed to make a plan of the Hutcheson lands in November 1789 and work began on a regular grid in 1794. Hutchesontown took longer to be established, however, partly as it relied on a link to the north bank of the river. The first bridge was washed away soon after construction in 1795 and was only replaced in 1804. Overall, the terms of the feus were set to create a uniform pattern of

[Charles Abercrombie], *Plan of Gorbal Barony* (c.1800)

four-storey residences along wide streets built within a set time.

Gorbals as an established settlement itself retained its village character, as can be seen from this early small plan prepared by Thomas Richardson. At the date of this survey, he was working from his map and stationer's shop in the Tron steeple. In 1797, a Council committee had inspected the lands, fences, buildings and marches in the village and its neighbourhood to ensure that there had been no encroachment by the Govan Coal Company. The following year, the intention to feu lands here was delayed until a survey and regular plan had been prepared but by 1802 the Windmill Croft lands were being auctioned in lots and a further area south of the Paisley Road was earmarked for new streets. The plan is, in effect, only an extract of the larger plan of the barony (p.54), coloured to identify the village and its boundary. Like the original, it is oriented with south to the top but omits the table of references. There are features unique to it, such as the windmill downriver from Jamaica Street Bridge, the parish burying ground, the toll on the Rutherglen Road and Shiel Loan.

In 1801, a separate part of the land to the west of the village was feued by the Hutcheson trustees to David and James Laurie. David soon put forward grand plans to develop the area as his own up-market suburb, Laurieston, with streets named after the English nobility. Both Gardner and David Smith are known to have inspected the lands in 1804. Unfortunately, Glasgow's expansion was to suffer from a shortage of wealthy potential property investors and the suburbs south of the river were no exception. Unlike other local speculators, Laurie lacked both the right kind of family or business connections and sufficient capital to ensure the complete realisation of his scheme. In addition, Laurie's intentions for the amenity of the new neighbourhood were soon to be affected detrimentally by the creation of a tramway built by the coal and iron master, William Dixon to link his Govan colliery with the Clyde. This was only the first of several intrusions fated to affect these districts as residential areas. Competition from both developments in the transport network and industrial concerns impacted on any sense of exclusivity. In 1814, Port

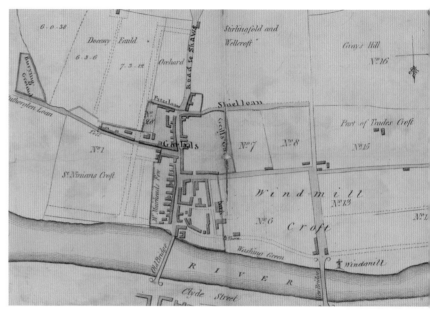

ABOVE. Thomas Richardson, *Plan of the Village & Parish of Gorbals* (1803)

OPPOSITE. Anon. *Plan of the Barony of Gorbals* (1850)

Eglinton was opened as a terminus for the Paisley and Ardrossan Canal and this would be followed by the Glasgow and Paisley Joint Railway line terminating at Bridge Street Station in 1840. The Govan ironworks at Dixon's Blazes were to become a constant reminder of the immediate proximity of industry.

In truth, the creation of higher-quality residential areas needed more than wishful thinking. Looking beyond the design of streets and houses, improvement came from the introduction of other facilities. In Laurie's case, his intentions to establish a town hall, public market and Grand Academy all came to nothing. Gradually, weaving factories and engineering works began to intrude into the residential pattern in Tradeston and Hutchesontown was also to see the establishment of cotton mills. These changes additionally created a demand for workers' houses and a gradual overcrowding. Sub-division of properties led to a decline and, despite the intervention of the City Improvement Trustees, the area was

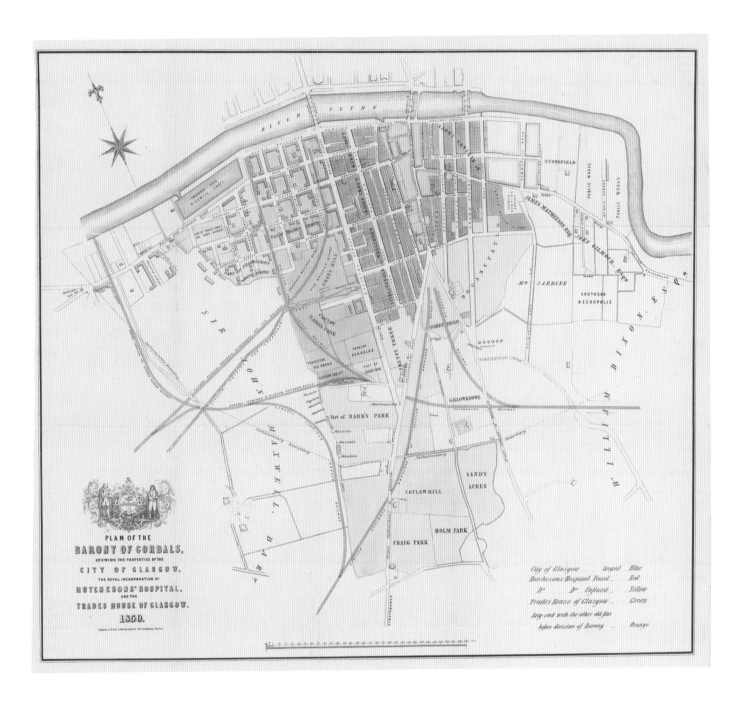

among the most congested districts in Glasgow until the 1960s. The solution of high-rise tower blocks did little to improve the area. In the last twenty years, however, many of these have been demolished and replaced with a more imaginative mix of private and social housing, enhanced by a better-quality landscaping.

1805 & c.1816

One estate, two plans: Possil at the beginning of the nineteenth century

While many estates were surveyed as part of a general process of agricultural improvement which saw rising rents and increased productivity, others were mapped when the sale of lands was involved. These two plans provide a 'before and after' picture of the Possil estate on the north side of Glasgow in the first decades of the nineteenth century. Like many other regional centres, Glasgow was surrounded by country houses which were within easy access of the city centre. During the seventeenth and eighteenth centuries, several were purchased by successful businessmen. Possil was no exception to this. Lying about three miles from Glasgow Cross, the two lands of Nether Possil were purchased separately in the 1740s by

Thomas Richardson, *Plan of the Estate of Possil* (1805)

59

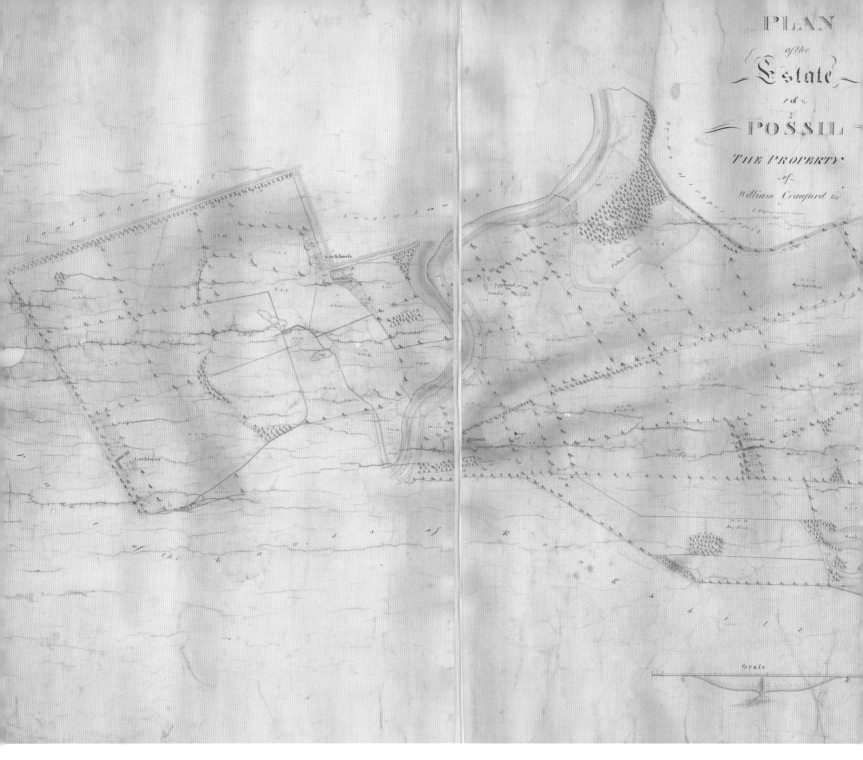

William Crawfurd of Birkhead, a Glasgow merchant. When his son Robert succeeded to the combined estate in 1772, he also came into possession of Langside, on the southern outskirts of the city. In 1805, Robert Crawfurd died and his son, William sold the estate to Colonel Alexander Campbell three years later. The new owner was the eldest son of John Campbell, the founder of one of Glasgow's most important companies trading with the West Indies. Their wealth was

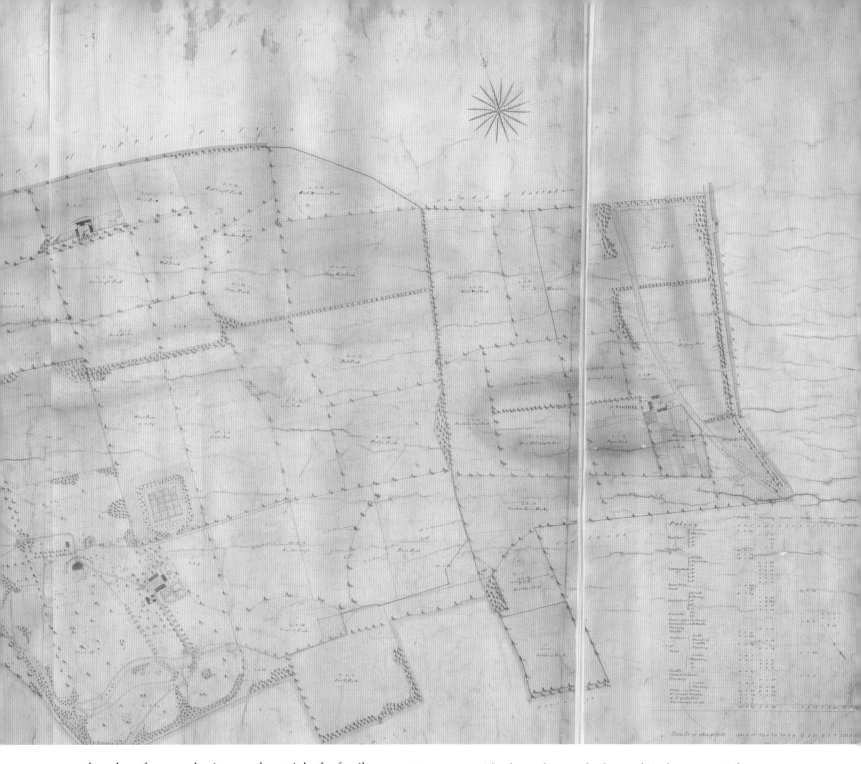

based on the sugar business and certainly the family were, at one time, slave owners. Alexander Campbell, however, was a distinguished officer in the British army and the year after buying Possil, served under Lieutenant-General Sir John

ABOVE. Richardson's plan was clearly a working document, as it has many pencil notations
OVERLEAF. John Yule, *Plan of the Estate of Possil* (c.1815)

PLAN

of

the ESTATE of

POSSIL

IN THE BARONY PARISH of GLASGOW.

And

COUNTY of LANARK.

The Property of

ALEXR. CAMPBELL ESQRE.

by

John Yule.

Scale of
1 Feet each?

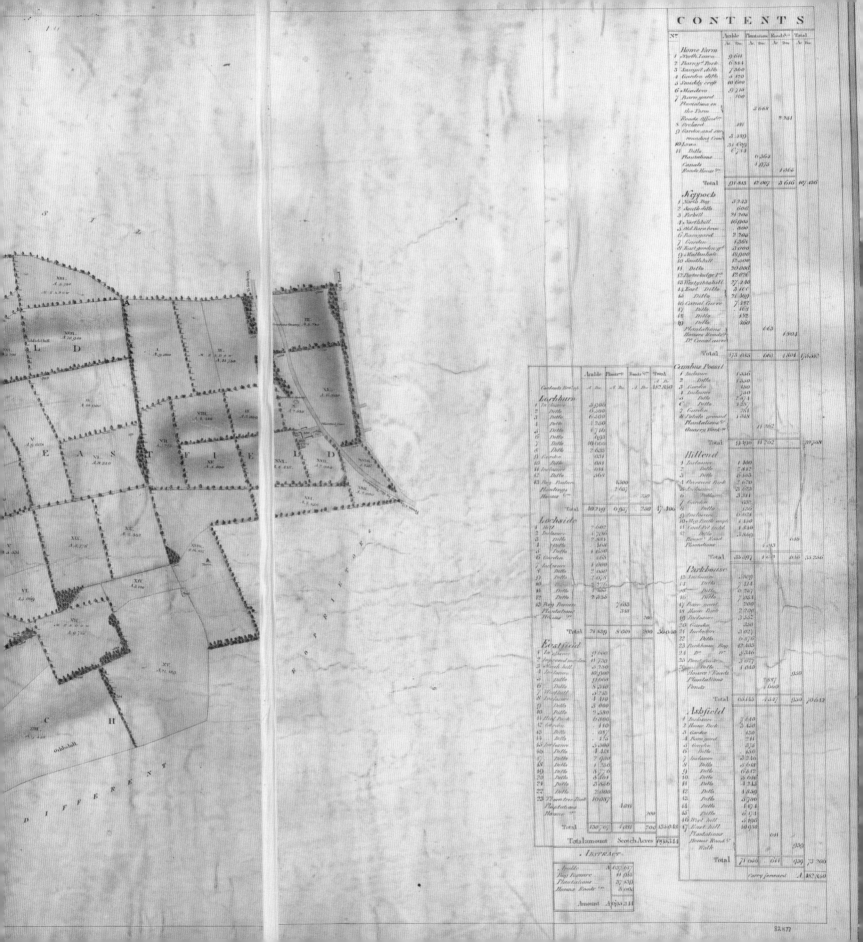

CONTENTS

Carry forward Ac. 182.850

82877

Moore, another local man, at the battle of Corunna.

The first plan was drawn by Thomas Richardson (1795) in 1805 and it can be assumed that it was commissioned when William Crawfurd inherited the lands. It is a fascinating document and, while some of the table of contents has been lost, it emphasises the importance of, and requirement for, a competence in mathematics to calculate the acreages of the constituent parts of the estate. This table details measurements for arable land, plantings, pasture, water, quarries and roads on the property's more than 200 hectares. It cannot be ascertained whether or not it was Crawfurd's intention, as the new owner, to introduce any improvements himself but, given the relatively short period of ownership, it would seem more likely that the plan was prepared as a document in support of the sale of the lands. However, it clearly has been well used and shows many signs of editing, with field names added and other pencil additions. These give increased value to it as an historical document. The whole area was under improvement – a plan of the neighbouring lands of Hamilton Hill and Keppoch had been drawn the year before by another Glasgow land surveyor, John Wilson.

While the second survey is undated, it is the work of John Yule, a relatively little known figure, who worked as an assistant to Robert Bauchop, factor to the Duke of Hamilton. Yule himself stated in 1815 that, with the exception of one year's training on the Dirleton estate in East Lothian, he was employed mainly on the Hamilton lands. He drew a map of the island of Arran, based on Bauchop's surveys carried out between 1807 and 1812, and, subsequently, prepared plans of the baronies of Lesmahagow and Dalserf up to 1815. By 1820, he had moved to Cumberland, where he was engaged as land agent for the Graham family at Netherby until the mid-1840s. It is difficult to date this Possil plan exactly but the quality of the penmanship underlines Yule's artistic skill, perhaps indicating that this was no apprentice piece of work.

Given his career history, it is most likely that this plan dates from the period after 1815 and before his move to England, at a time when the end of the Napoleonic Wars would usher in major changes in agricultural society.

On purchasing Possil, Campbell made many additions and improvements to the house and its policies. At that time 'it was then far away from the noise and smoke of the city, and stood among fine old trees. With its beautiful gardens, its grassy slopes, and its clear lake, Possil formed as delightful and retired a country residence as any in the county'. Although only a few years separate the production of these plans, alterations can be seen in the layout of the grounds immediately adjacent to the house, particularly the removal of the home farm to the new offices north of the formal garden. In Yule's plan, the estate boasts a 'melonry', the plants being kept at a suitable temperature from the heat produced from the nearby sawpits. Elsewhere, new ponds, a smithy and a vitriol works are marked and, more significantly, the estate has grown to nearly 285 hectares.

Campbell subsequently acquired the neighbouring property of Keppoch, in addition to the estates of Torosay and Achnacroish on Mull, but Possil did not remain the family residence. Coal and ironstone were later found in the area. Parts of the estate, including the house and immediate gardens, were feued to the iron founder Walter MacFarlane who replaced them with the Saracen foundry and renamed the estate Possilpark. Between 1872 and 1891, the population grew from 10 people to 10,000 and a new suburb of tenements designed on a grid plan was created. With the closing of the foundry in the late 1960s, the name Possilpark became associated with crime and serious social problems. According to the 2012 Index of Multiple Deprivation, it was the second most deprived area in Scotland. Since the late 1990s, the whole area has been undergoing mass redevelopment and there has been much investment in local services.

OPPOSITE. The re-designed offices and melonry of Yule's plan

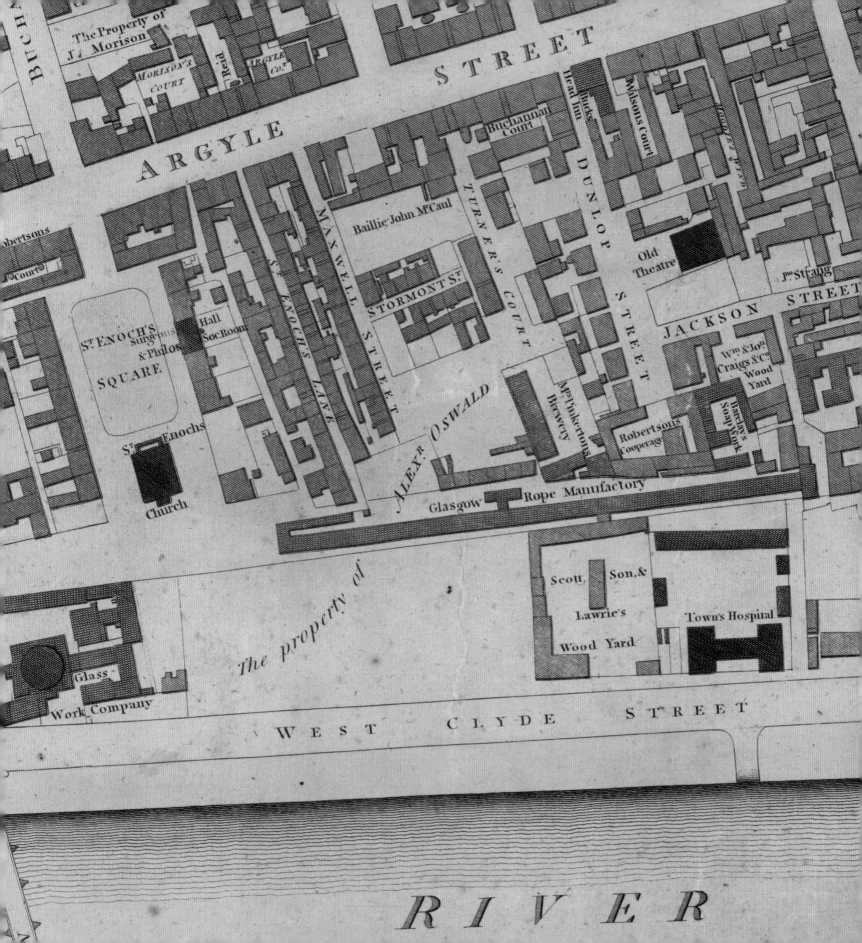

ARGYLE STREET

The Property of J. Morison

MORISON'S COURT
Reid
ARGYLE COT

BUCHANAN
Buchannan Court
Head Inn
Ducks
DUNLOP STREET
Wilsons Court
GOOSE DUBS

Robertsons
Court

Baillie John McCaul

TURNER'S COURT

Old Theatre

J°. Strang

JACKSON STREET

ST ENOCH'S
Surgeons
& Philos.
Hall Soc Room

St ENOCH'S LANE

MAXWELL STREET

STORMONT ST

SQUARE

Wm & Jon
Craigs & Co
Wood Yard

Harlays
Soap Work

St Enochs

ALEXr OSWALD

Mrs Thakertons
Brewery

Robertsons
Cooperage

Church

Glasgow   Rope Manufactory

Scott, Son, &

The property of

Lawrie's

Town's Hospital

Glass

Wood Yard

Work Company

WEST CLYDE STREET

RIVER

1807

'Executed in the neatest manner by a respectable Artist': Fleming's map of Glasgow

Peter Fleming's name first appears in the Glasgow press in 1802, announcing an intention to start business in surveying estates and the accurate measurement of grounds. Once again, this emphasises the importance of competent mathematical skills and indicates the type of work surveyors were called upon to provide. Early in his career, he shared an address with William Kyle, founder of a school of Glasgow surveyors at the beginning of the nineteenth century. It is most likely that Fleming trained under him and went into brief partnership with his mentor in 1804. This was an ideal time for someone of his ability to be working in Glasgow. Between 1801 and 1821, the population very nearly doubled. City growth, new suburban projects and a range of transport developments all contributed to a demand for the talents of a skilled measurer. This is reflected in the surviving body of work Fleming produced, which includes street maps for the Council, a survey of New Lanark and a report on lowering the level of Loch Lomond, as well as several industrial, estate and property plans. These also show the gradual transformation of his career from surveyor to civil engineer, which was representative of an increasing specialisation of the profession overall.

Fleming was a man with big ideas and a determination to produce reliable work. At this date, the publication of a six-sheet map of the city was a large undertaking and, while the title indicates its sale by a Glasgow bookseller, the engraving was done in Edinburgh by Robert Scott (1797). The advertisement for the plan emphasises its basis on actual survey but this was an expensive piece of work. It would cost the purchaser two guineas coloured and another eight shillings if mounted on rollers and varnished. Nonetheless, only two months after this newspaper announcement, the Council resolved to obtain copies, coloured to distinguish the individual parishes, to present to each city minister. In April 1808, Fleming purchased his entry as a burgess and guild

Peter Fleming, *Map of the City of Glasgow and Suburbs* (1807)

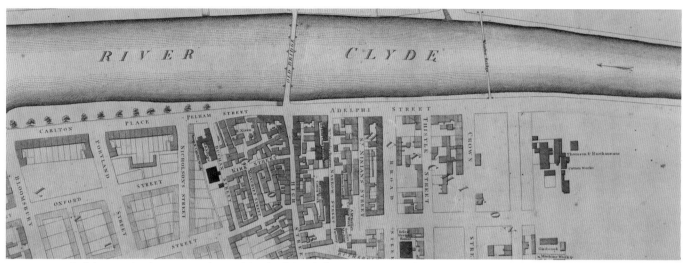

ABOVE. Detail of the new development on the south bank of the Clyde

OPPOSITE. Fleming's plan marks what are, in reality, intended street patterns in Blythswood and Tradeston

brother of the city and followed this up with a failed subscription scheme to produce a map of the low counties of Scotland. Seven years later, he published *A System of Land Surveying and Levelling*, intended as a complete treatise on measured chain survey, with an emphasis on the importance of geometry in calculations. In his conclusion, he stressed that 'sometimes a bad practice with a painted plan, or low charges, is preferred to true results, which preference must alone arise from not observing that all methods of measuring land have not alike verifications'.

The immediate impact of this map is its depiction of the spread of the city both to the east and west, particularly in the designs for new streets on Archibald Campbell's Blythswood estate and the westward extension of Tradeston and Laurieston. Although these are delineated in a strict grid fashion, the blocks are mostly lightly shaded, with only the occasional structure in a darker tone. All of this underlines the indefinite nature of what were, in effect, proposals at a time of rapid change in the structure of the city. This is especially true with the angled intrusion of Alston Street running north from Anderston. Equally important is Fleming's detail of the industries developing in and around the city.

Many quarries, works and factories are identified, along with the names of their owners, for example, the dye and cotton works off Duke Street. A degree of industrial specialisation in particular districts can be identified. In Calton and Bridgeton, nine cotton works and mills are located. Other notable features include tolls, the Nelson monument on Glasgow Green (Britain's first civic monument erected to his memory in 1806), public buildings, hotels, places of worship and the two canal systems lying to the north. As a member of the Glasgow Philosophical Society, Fleming was canny enough to indicate its meeting rooms in St Enoch Square.

Although the title design is illustrated with images of three sailing vessels, as well as a sketch of a wharf, three mariners and a small boat named 'Clyde', this allusion to Glasgow's trade is not repeated on the map itself, where there is no indication of ships on the river and only the Broomielaw quay marked. In addition, while there is no hill shading or hachuring to indicate the shape of the ground, spot heights are given. An unattributed copy of McArthur's plan of 1779 is included in the south-west corner, possibly to underline the rapidity of change but also to place Fleming's own work on a par with that of his predecessor. In 1808, Fleming followed

McArthur's example further by issuing a single-sheet reduction, and these two maps were to have a profound influence on all subsequent city plans up to at least 1848.

Sometime after June 1823, Fleming emigrated to the United States, where he worked on the construction of the Mohawk and Hudson Railway before moving to Canada.

Here, he was employed as engineer on a variety of bridge, canal and river projects and was an early promoter of a railway link from the St Lawrence River to the Great Lakes. By the mid-1840s, he had fallen out of favour with his employers, possibly because of a contentious nature, and turned to writing mathematical works.

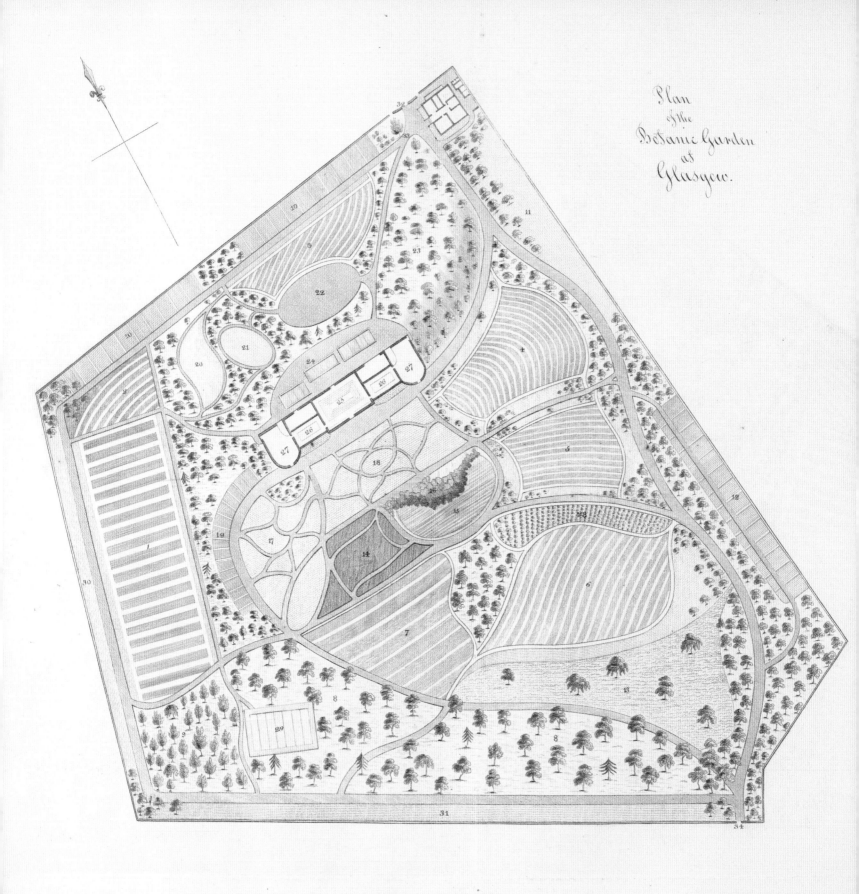

Plan
of the
Botanic Garden
at
Glasgow.

1818

'That plot of ground which is dignified with the name of Botanic Garden'

As early as 1704, the University laid out a physic garden in the College grounds to provide plant specimens for study as part of the drive to establish medical education there. Fifty years later, Professor William Cullen was commenting that the plants are 'very much exposed to the smoke and soot of the Town'. By the beginning of the nineteenth century, the professors were particularly conscious of the impact of the poor clayey soil and the heavy metal poisoning coming from the type foundry located in the University's own grounds. Declining productivity and the generally poor conditions for plants in the environs of the High Street site stimulated the intention to find a suitably located alternative. After selling some of the garden ground in 1813, it was initially planned to create a new garden at Blythswood. In that same year, Thomas Hopkirk of Dalbeth (1785–1841), a pioneer local botanist and Fellow of the Linnean Society, published his *Flora Glottiana,* one of Britain's earliest regional floras, which catalogued over 100 species indigenous to the Glasgow neighbourhood.

Three years later, Hopkirk, aware of the city's lack of such a resource, was largely instrumental in forming a society intent on establishing a Botanic Garden to display plant specimens, including his own extensive collection. This group quickly gained support from several leading local figures, including James Smith of Jordanhill and James Ewing, later to become Lord Provost and M.P. for the city. Within a short time, over £6,000 was raised by the issuing of shares. In addition, the University authorities, seeing this as an answer to their own immediate problem, agreed to contribute a further £2,000 on the understanding that its students would have access to plant specimens and that the Professor of Botany would deliver a series of open lectures each summer. These deliberations were to receive Royal acknowledgment in a Charter issued in September 1818 for the creation of the Royal Botanic Institu-

---

Anon. *Plan of the Botanic Garden at Glasgow* (1818)

tion of Glasgow. Additionally, the Crown paid £2,000 towards its foundation.

The Garden opened in 1819 with 8,000 plants on a site covering more than 3 hectares at Sandyford, then outside the burgh boundary, and was another symbol of the city's westward expansion. A lecture room which could hold 200 people was built within the grounds. Two years earlier, while negotiations were going on to establish the garden, the Regius Chair of Botany at the University of Glasgow was established – an indication of the importance of the subject in the teaching of medicine. The first occupant of the chair was Robert Graham but, in December 1819, he was appointed Professor of Medicine and Botany at Edinburgh University, subsequently becoming deeply involved in planning the new botanic garden at Inverleith. Following advice from Sir Joseph Banks, president of the Royal Society, his successor at Glasgow was William Jackson Hooker (1785–1865), later knighted for his services to botany in Glasgow and first Director of Kew Gardens. He was destined to make Glasgow a major centre for the study of plants and prepared a catalogue of the Garden's more than 12,000 species and varieties. It was under his skilful guidance that the Garden achieved its remarkable success and international stature. In 1821, he published *Flora Scotica*, a description of Scottish plants based on his extensive field trips. His students included David Douglas (1799–1834), one of the country's most outstanding plant collectors.

FASHIONABLE PROMENADE.

ROYAL BOTANIC GARDEN.

This fashionable place of resort has this season been more numerously attended than ever. We observe that additional grass walks have been added to the Promenade ground, and the excellent Trumpet Band of the 5th Dragoons, we believe, will be in attendance on Saturday next. The splendid Cactus Speciosissimus which has been in flower for some time past, has given increased interest to the Houses. ———— Vide et crede.

OPPOSITE. Diploma presented to honorary members of the Royal Botanic Institution of Glasgow

ABOVE. Illustration of the Garden from *The Glasgow Looking Glass* (1825)

This plan first appeared in the *Companion to the Glasgow Botanic Garden, or Popular Notices of the More Remarkable Plants Contained in it*, published about 1818. Contemporary comments remarked on the combination of variation of landscape with scientific arrangement. Numbers on the plan refer to the various plants, borders and other components. The layout of the garden was largely the design of Stewart Murray, Hopkirk's own gardener, who was appointed to the post of Curator. Features of this scheme, particularly the winding pathways, were to re-appear in the plan of the Glasgow Necropolis, opened in 1832 as Scotland's first garden cemetery, on the rocky slopes of the Fir Park lands. The Sandyford location quickly became a popular resort for promenaders and regiments stationed in Glasgow would send their bands there to entertain visitors. By the late 1830s, there were numerous buildings surrounding the Garden, preventing its expansion and, with the collection of plants growing, a new site was needed. In November 1839, the Institution entered into possession of its present site on Great Western Road, where Joseph Paxton, later knighted for his revolutionary design for the Great Exhibition of 1851, was commissioned to lay out the garden on lands on the south bank of the River Kelvin.

Hopkirk was a talented illustrator, becoming editor of *The Glasgow Looking Glass* in 1825, in which the new Botanic Garden was portrayed. References to him suggest that he was responsible for introducing lithography to the city but he is not recorded in the major printing directories. He subsequently moved to Ireland and was involved for some time with the geological department of the Irish Ordnance Survey. Both the University and the Institution recognised Hopkirk's contribution to botany by naming the Hopkirk Laboratory for taxonomic biology and the Hopkirk building in the Botanic Garden respectively in his memory.

1820

Mapping a rural estate: Smith's plan of Milngavie village

Milngavie lies seven miles north of Glasgow's city centre and is now a popular part of the suburban commuter fringe of the city. This plan brings together several of the threads of Glasgow's history which have been discussed already. While it is clearly another estate plan covering lands owned by one of the commercially successful Glasgow merchant families, it also indicates the development of an industrial base in what was a more rural location.

It was prepared by David Smith relatively early in his career, in 1820, for the new owner of the estate of Dougalston, a member of the Glassford family. James Glassford (1771–1845) was a Scottish legal writer and traveller who was admitted to the Faculty of Advocates in 1793. He was one of the first to treat legal evidence as a distinct subject, advancing a more complete approach to the evaluation of testimony. A recognised scholar, he translated the Latin writings of Francis Bacon and the work of several Italian poets, including Ludovico Ariosto. As well as being a member of the Highland and Agricultural Society, he served as sheriff-depute of Dunbartonshire between 1805 and 1815. His father, John Glassford, was a leading Glasgow tobacco merchant and at one time Scotland's greatest ship-owner, whose sizeable property, Shawfield Mansion, in the newly developing lands north of Trongate is clearly marked on John McArthur's plan (1778). In 1819, James succeeded to the Dougalston property on the death of his elder brother Henry and, in the close-knit network which was the Glasgow mercantile elite, it is interesting to note that their sister, Christian, was the mother of Thomas Hopkirk, founder of the Royal Botanic Garden.

Smith's depiction of both the estate and the village of Milngavie was obviously drawn soon after Glassford entered into possession. It is one of a series, which included a volume of farm plans based on work done in 1805, which he prepared in the March of that year. Drawn in more detail than those

---

David Smith, *Plan of the Village of Milngavie, with sundry adjoining pendicles* (1820)

contemporary copies, the Milngavie plan is also the first of a sequence of maps of the lands that Glassford had prepared by several local surveyors, who also included James Shanks, up until 1844. While David Smith is, arguably, better known for his several plans of Glasgow itself, this work shows clearly his artistry and ability to produce an accurate record. Properties are numbered and marked alphabetically, while land use is also indicated. Neighbouring estates, including Glassford's own land of Clober, are identified and a sense of the area's relief is provided by a grey colour wash.

Of possibly greater significance is Smith's indication of the changes which were already happening in the immediate area. An apparently more residential new town is marked well to the east of the original settlement, on the turnpike road running between Glasgow and Buchlyvie. More importantly, Smith shows the extensive grounds held in long lease by the Milngavie Cotton Spinning Company, along with its mills, bleaching greens and reservoirs beside the Allander Water.

It was this water supply which encouraged the development of industry in the immediate neighbourhood. Water was to have an even greater subsequent importance to the area when the reservoir at Mugdock was created in the 1850s as part of the supply system drawn from Loch Katrine. Through his second marriage, to the daughter of James McGregor, a Glasgow bleacher, James Watt became involved in the development of bleaching at Clober by laying out a bleachfield, designing water courses for power and installing machinery.

During the nineteenth century Milngavie grew from a country village to a small but locally important industrial centre. Cotton spinning had been established in the area in 1790 by Henry Glassford with the erection of a mill which attracted several workers from Perthshire. As with Glasgow itself, the development of manufacturing led to an increase in population. According to the report written by Rev. George Sym for the original *Statistical Account* in 1793, Milngavie had about 200 residents. By the time his son Andrew prepared the parish entry for the *New Statistical Account* forty years later, the population had risen to 1,162. Sym directly attributed this almost six-fold increase to the cotton factory and calico

print works, enhancing the town's local reputation. Other businesses were attracted to the area, for employment was also provided by a paper, a snuff and four corn mills, in addition to a distillery.

By the 1830s, Andrew Sym's record was describing the introduction of machinery to much of the manufacturing process, as well as a strike by the united printers, while, in the same year as the preparation of Smith's plan, Glassford's estate factor Henry Gordon reported a failed attempt to set fire to the cotton mill by two men. This was part of a growing industrial unrest caused by falling prices, reduced wages and the increased employment of women at lower rates. Eventually, cotton spinning declined as a result of the loss of a regular supply during the period of the American Civil War. Surviving correspondence between Gordon and Glassford shows a typically commercial approach to the management of the estate, characterised by a frequent discussion of leases. However, a more charitable aspect of business is suggested in a letter sent in December 1835, where Gordon refers to his employer's offer to provide equipment for the newly opened infant school.

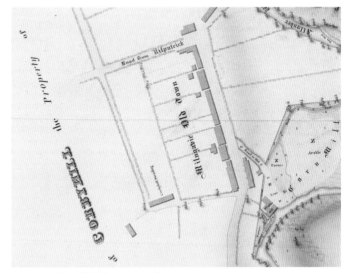

ABOVE. The old Milngavie settlement
OPPOSITE. Smith's plan marks the cotton mills located beside the water supply of the Allander Water

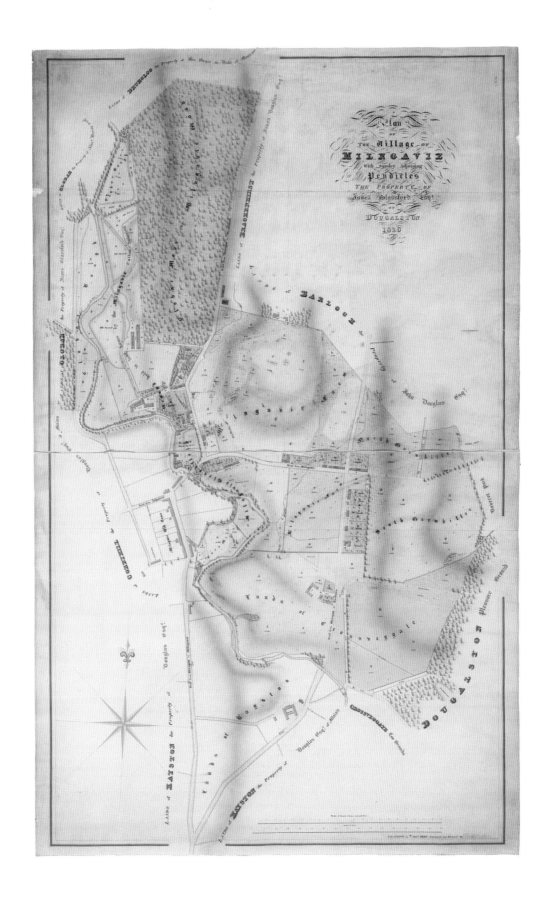

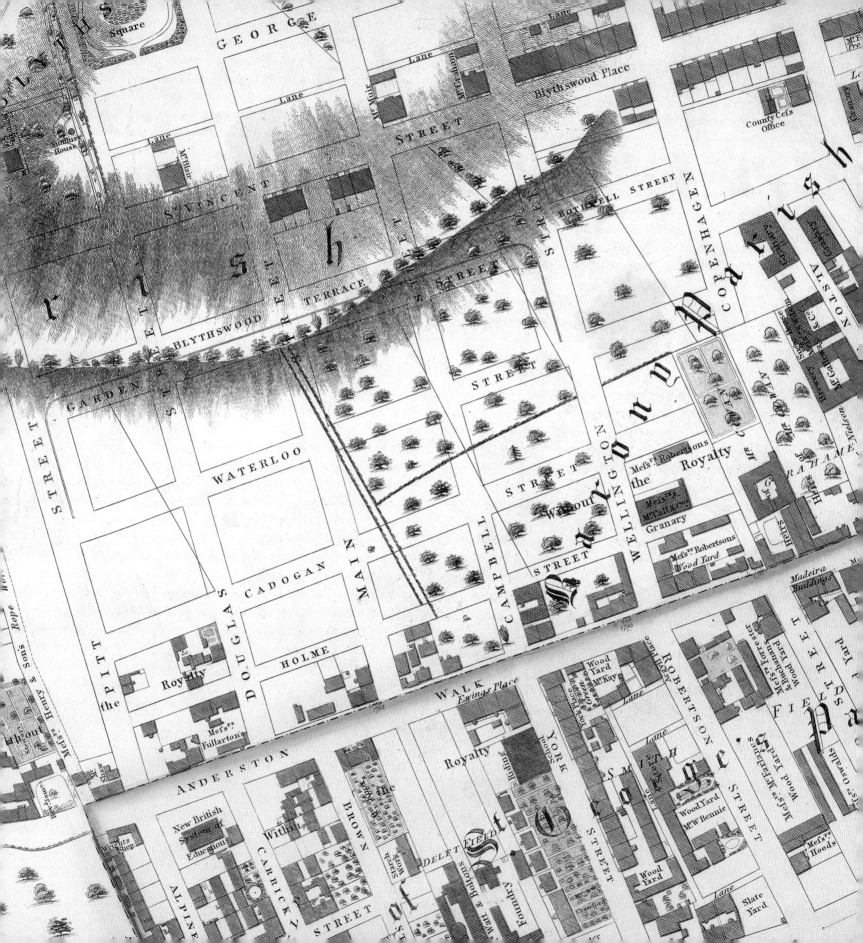

1821

'More complete than any thing of the kind yet done': a 'new' map of Glasgow

The departure of Peter Fleming to North America in the early 1820s may have left a void in the ranks of the local surveyors but it was well filled, possibly even before he crossed the Atlantic. Another of William Kyle's pupils, David Smith, was already making his mark. He is first recorded in 1804 and his earliest known address was in Double Dykes, Calton in 1811. The plan of the Dougalston lands (1820) and a carefully executed manuscript map of the royalty of the city of the same year are early examples of his work. In March 1821, Fleming issued proposals for what would have been the most ambitious mapping project for the city before the arrival of the Ordnance Survey in the 1850s. His scheme was to produce a map of Glasgow at a scale of one inch to forty feet (1:480), showing every building and all the proposed alterations, along with a list of proprietors' names. The public were asked to subscribe at 7/6d per sheet (or 5/- each, if committing to the complete work) and it was to be prepared using the most modern technology, including lithographic printing. Fleming's advertisement hints at the rapid growth of the city over the intervening period for he mentions the limitations of his 1807 map for determining accurate dimensions, particularly with respect to projected improvements. A fortnight later a further notice appeared, noting that the first sheet would be published in a few weeks, but it was to be the last the citizens heard of this uniquely original proposal.

Almost a year to the day following that final notice, a new six-sheet map of the city, published by Alexander Finlay and William Turnbull, a local bookseller, was advertised in the *Glasgow Herald*. On the same page Turnbull also announced his publication of a translation of Charles Nodier's *Promenade from Dieppe to the Mountains of Scotland*. Based in the Trongate, Turnbull was in partnership with Finlay (c.1774–1825), a carver, gilder and looking-glass manufacturer, who offered for sale collections of English and foreign prints as

David Smith, *Map of the City of Glasgow and Suburbs* (1821)

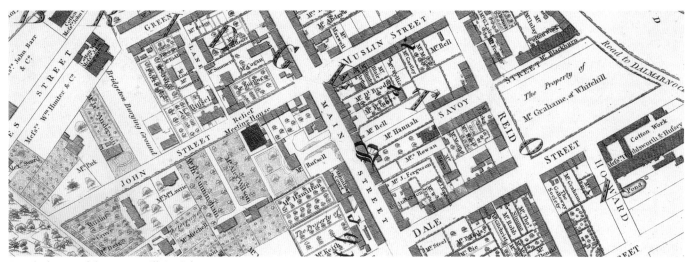

Detail of some of the cotton mills in Calton

well as drawing materials. His business developed into advertising engravings and, by 1815, he described himself as 'Print-seller to His Majesty'. In 1821, he opened a new picture gallery in South Maxwell Street and played a significant role in the foundation of the short-lived Glasgow Institution for Promoting and Encouraging the Fine Arts.

Surprisingly, the press notice mentions neither Fleming nor Smith in connection with the map. Nonetheless, this is the map which they offered for sale, highlighting that the alterations had been brought down to Whitsunday 1821. What is uncertain is both the level of responsibility that either partner had for the new survey and how they procured the original plates. The link between Fleming and Smith through William Kyle cannot be overlooked but the exclusion of any mention of Fleming, combined with the lack of any further local cartographic evidence, suggests that he had nothing to do with this revision. On the other hand, Smith fully recognises that the plan is based on Fleming's original work. A title note confirms this and records Smith's own contribution as 'additional Surveying for laying down the extension of the City & Suburbs'. The map certainly reflects this, for the re-use of the engraved plates results in the same area of coverage, the same style and positioning of lettering and the same scale. This is,

however, no mere copy for Smith has recorded, in meticulous detail, the significant additions and alterations to the 'pattern' of the city in the intervening fourteen years.

This was a period of rapid population growth and, in many areas, Smith's changes amount to a virtual re-drawing of the original survey. This is particularly true in the street system in the Garnethill and Blythswood area to the north-west, where the grid layout of the district is strikingly different. Premises are now indicated north of Anderston Walk and, while there are still relatively few houses marked, the design of Cadogan, Copenhagen, Wellington and Bothwell Streets appears slightly less indefinite in intention. Some of this remains tentative and the pecked delineation of St Vincent Street underlines that not all the features mapped existed. Attention has been given to the indication of gardens, pleasure grounds and footpaths. In addition to many new buildings and features, Smith has taken pains to introduce hachures and hatching to indicate both the topography of the area and the city's several quarries.

More definite change and growth appears to be mapped on the eastern fringes of the city in Bridgeton, especially in the buildings identified in the area around Muslin, Dale and Reid Streets. At this date, hand-loom weaving was predominant in

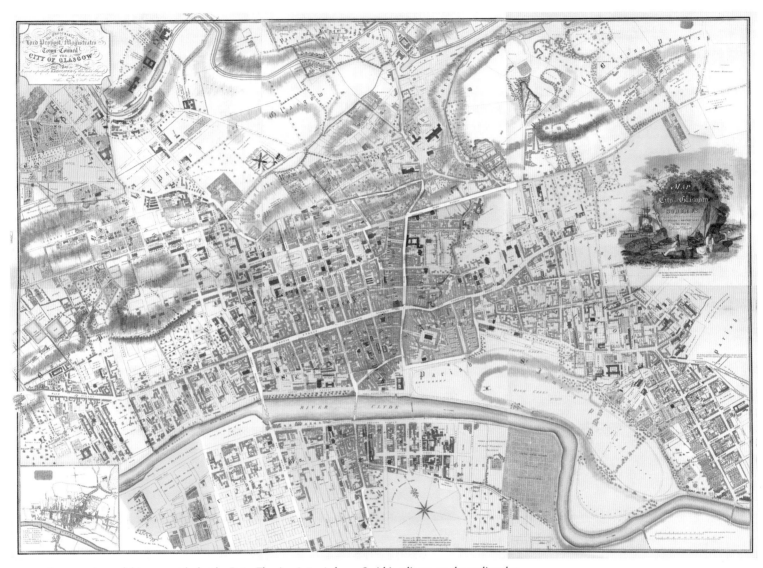

A comparison of this map with that by Peter Fleming (1807) shows Smith's reliance on the earlier plan

the area and, while two cotton works are shown on James Street, much of the industry was still carried out in the workers' own homes. The map's greatest value may be in the increased detail of the city's industrial concerns where property names and usage are recorded. New transport developments can be seen in the growth of the harbour downstream of the Broomielaw, which included a tracking path and basin at Windmill Croft. Additionally, as a sign of things to come, a rail road and coal store is located in Barrowfield. Less obvious but equally important is the naming of parishes and police wards, reflecting the growth in the administration of the city. One final change should be noted in the title cartouche, where a steam vessel has been added and a more abundant foliage design is a possible allusion to the city's healthy prosperity.

1822

A map for a census

This plan is generally known as 'Cleland's map' and it is proper to record his association with the city he served so diligently. James Cleland (1770–1840) was a statistician and superintendent of public works for the burgh from 1814 to 1834. It was while he held this latter post that he was to have his greatest influence on the residents of Glasgow. During his time in office, his projected works included a fruit market, a cattle market and new bridges across the Clyde. In addition, he was responsible for a number of Glasgow's churches to cater to the growing population, including St David's, now known as the Ramshorn Kirk, and the redevelopment of Glasgow Green. He also played an important role in the introduction and standardisation of the city's weights and measures.

His name, however, is indelibly linked with his pioneering work on population enumeration, where his flair for statistics came to the fore. In 1819, on the Council's behalf, he undertook the most extensive and detailed local census that had ever been conducted in Britain. Several of his innovations in demographic statistics were adopted subsequently by the government for the national censuses of 1821 and 1831. In an effort to determine life expectancy, he prepared bills of mortality for the city between 1820 and 1834. Cleland was in the van of a growing body of interest in the collection of statistics on the structure of society. He corresponded with Sir John Sinclair, contributed to both the Glasgow and Rutherglen *Statistical Accounts* and was a member of many learned societies. In recognition of his services to the city, a subscription was instituted and a building at the north end of Buchanan Street was erected, designated 'The Cleland Testimonial'. Its pediment can still be seen clearly from the steps of the Royal Concert Hall.

Dedicated to the Lord Provost, the title emphasises this map's indication of the ten city parishes and that it was prepared specifically for the census. While Cleland's name

David Smith, *Map of the Ten Parishes within the Royalty and the Parishes of Gorbals Barony of Glasgow* (1822)

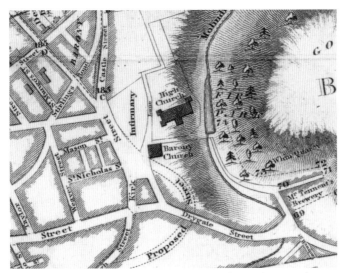

ABOVE. An example of new road proposals and an indication of Tennent's Brewery

OPPOSITE. The whole map displays the considerable detail of the administrative jurisdictions and boundaries of the city and the neighbouring lands

has been no great effort to identify businesses, offices or industry. In fact, few individual buildings are identified.

North of the river, Smith tries to show the drumlins on which the spreading city was being built. Hachuring is used to indicate relief and several local hills are named. New features appearing include the Magdalene and lunatic asylums off Dobbies Loan. Unusually, while there was no indication of it on the earlier Smith city plan (1821), the Royal Botanic Garden is now named. More significantly, there is a far wider and more detailed coverage of the lands south of the Clyde, stretching as far as Strathbungo. Smith marks the Paisley and Ardrossan Canal, along with its basin and a railway running south-east from it. It is also in the Gorbals that property ownership details are most obvious. Cleland had been appointed chief magistrate of that barony in 1804 and, like Blythswood, it was an area of rapid housing development.

Around the edges of the mapped area there are several notes relating to the city's administrative jurisdiction and that of its neighbouring lands. This attention to a more functional purpose behind the map carries over to the addition of a smaller-scale inset depiction of the royalty of the city, as well as a careful delineation of that boundary and its march stones. This map was also to appear in John Wood's town atlas of Scotland. Although this was untitled and undated, the accompanying *Descriptive Account of the Principal Towns in Scotland*, published in 1828, indicates clearly that Wood issued and collected the plans to suggest further urban improvement. Along with several other town plans, such as those of Leith, Paisley and Greenock, it shows Wood's reliance on local mapping and brings into question his own contribution. His descriptive account of Glasgow, however, runs to over twenty pages and focuses on the city's trade, industries and exports. More significantly, Wood refers his readers to Cleland's own work, *Annals of Glasgow*, for more information and appends his statistics to the text. In 1831, a second state of the map with certain additions (e.g. Garnkirk railway line, South Quay) was issued to accompany the second edition of Cleland's *Enumeration of the Inhabitants of the City*, published the following year.

appears prominently, an engraver's note states that it 'is constructed ... from a number of other detached maps & plans of acknowledged accuracy by David Smith'. Certainly, a manuscript draft by Smith exists, dated 1820 and possibly based on a perambulation of the city's marches in 1817, suggesting that this engraved example supports Cleland's own enumeration rather than the national census. It was first advertised in April 1822, priced at 15/- plain or one guinea if coloured and on rollers. Additionally, this notice repeats the map's own statement underlining the careful scrutiny behind its accuracy.

Cleland had been involved in dividing the city into parishes and, not surprisingly, the map indicates parish churches as well as delineating their boundaries. The majority of the city's streets are named and some proposed developments shown, most notably at the east end of Trongate, where two projected streets of seventy-feet width lying either side of Gallowgate extend from the Cross. Conversely, apart from Shawfield Printfield, Mr Tennent's Brewery and the waterworks, there

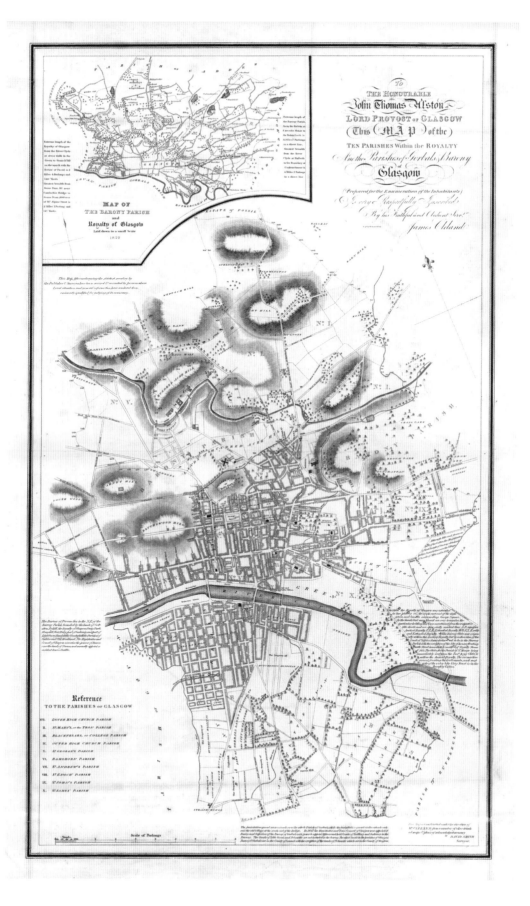

To THE HONOURABLE
John Thomas Alston
LORD PROVOST OF GLASGOW
This (MAP) of the
TEN PARISHES Within the ROYALTY
And the Parishes of Gorbals & Barony
of
Glasgow

Prepared for the Enumeration of the Inhabitants
Is very Respectfully Inscribed
By his Faithful and Obedient Servant
James Cleland

MAP OF
THE BARONY PARISH
and
Royalty of Glasgow
Laid down to a small Scale
1827

Reference
TO THE PARISHES OF GLASGOW

N°.	
I.	INNER HIGH CHURCH PARISH
II.	ST. MARY'S, or the TRON PARISH
III.	BLACKFRIARS, or COLLEGE PARISH
IV.	OUTER HIGH CHURCH PARISH
V.	ST. GEORGE'S PARISH
VI.	RAMSHORN PARISH
VII.	ST. ANDREW'S PARISH
VIII.	ST. ENOCH PARISH
IX.	ST. JOHN'S PARISH
X.	ST. JAMES' PARISH

Scale of Furlongs

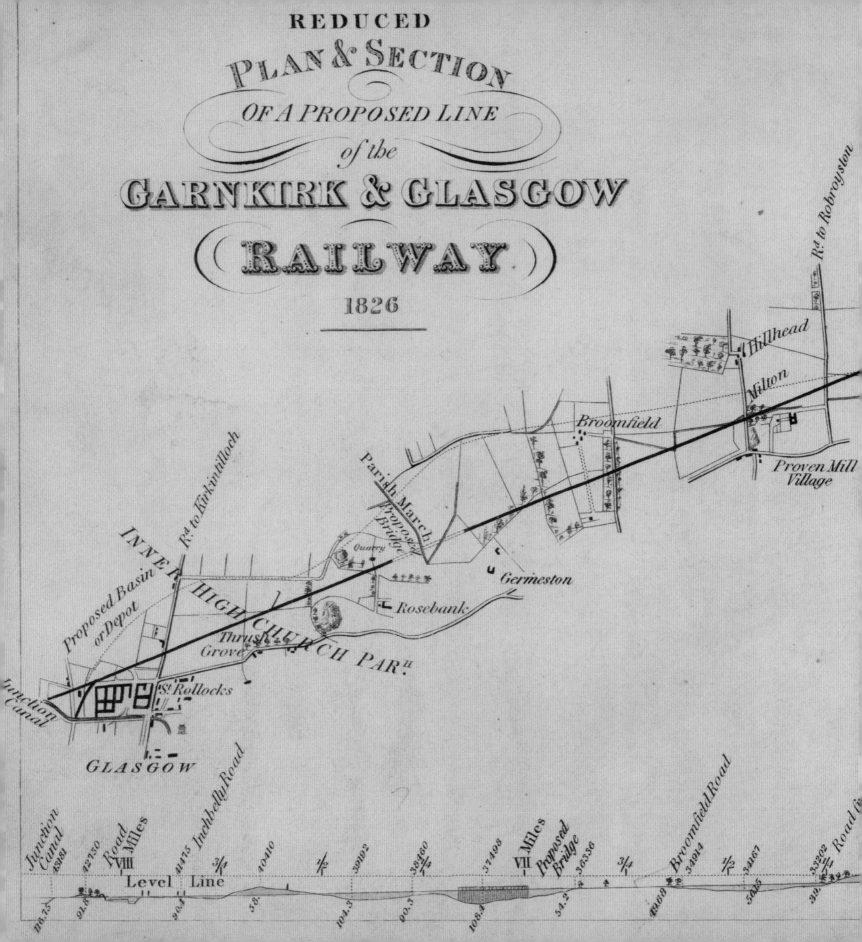

1826

The Garnkirk and Glasgow Railway:
Glasgow's first passenger line

This proposed plan and section was presented to the promoters by the leading railway engineers of the day, Thomas Grainger and John Miller, in 1826. Grainger had been apprenticed to the Edinburgh land surveyor, John Leslie, before setting up in business for himself in 1816. Miller was taken into partnership the year before this proposal and together they designed much of the early railway network north of the Border. Contemporaneously with this scheme, they had been working on their first rail project, the Monkland and Kirkintilloch Railway, opened in that same year of 1826 and designed to carry coal from the colliery at Cairnhill to the Monkland Canal. The partnership was also involved in a number of other local lines, including the Ballochney, Wishaw and Coltness, and Pollok and Govan Railways.

The first railway in Scotland authorised by act of parliament was the Kilmarnock and Troon, constructed to move coal from the Duke of Portland's pits to navigable water.

Opened in 1811, it was designed for horse-drawn wagons but problems with the iron tram plates meant that it was unsuitable for a locomotive. Nonetheless, a growing body of opinion realised the importance of the railway not merely as an adjunct to the canal system but as an entirely separate transport network in its own right. The future was seen in the development of the steam locomotive engine and, as reported in the *Glasgow Herald* in November 1824, this invention was regarded as 'nearly perfect in its construction, and it is efficient almost beyond belief in its operation'.

As elsewhere, industrialists, rather than those involved in transport, took the lead in supporting most of the railway lines in North Lanarkshire. Much of the demand behind the promotion of the Garnkirk and Glasgow scheme was very similar to that for the Monkland Canal, namely the need to transport cheap coal, iron ore and other minerals directly to Glasgow. The line sought to undercut the high prices being

Thomas Grainger & John Miller, *Reduced Plan & Section of a Proposed Line of the Garnkirk & Glasgow Railway* (1826)

charged by both the coal owners and the canal company and it was the first rail route to enter the city. One of the leading sponsors was Charles Tennant, whose thriving St Rollox chemical works at Townhead required a greater supply of materials than could be handled by the canal. Tennant was an enthusiastic supporter of railways and, along with Thomas Grainger, visited the north of England to inspect the developing network there. Royal assent for the Garnkirk and Glasgow Railway bill was received in May 1826 but an amended act the following year authorised an alteration in the course of the route.

Like its neighbours, the length of the railway was relatively short, running 13.2 km to link the Monkland and Kirkintilloch Railway at Gartsherrie with a depot in Glasgow. This terminus was to be located beside the cut of junction linking the Monkland and Forth and Clyde Canals but, more importantly, it was immediately adjacent to Tennant's Townhead works. This station remained the only one in Glasgow for ten years, inconveniently located on the city's fringe. Supervised by both Grainger and Miller, work began on constructing the line in August 1827. As it was intended to carry heavy goods and designed to use locomotives, the route was planned with gentle gradients, but with a notably heavier track construction. This required a considerable amount of work on embankments, cuttings and, particularly, laying the bed across Robroyston Moss. Bad weather further delayed progress and additional capital had to be raised but in May 1831 horse-drawn coal trains to Glasgow started to operate.

By the end of June 1831, the first locomotive, St *Rollox*, was in service. Three months later, a ceremonial opening took place, with inaugural passenger journeys pulled by steam engines in both directions, supervised again by Grainger and Miller. Uniquely, the day was recorded by David Octavius Hill in three lithographs depicting both the trains and the line. Reduced carriage rates, introduced on opening, were a direct challenge to both the Canal and the Monkland and Kirkintilloch Railway. The line's success, however, was largely due to its convenience, regularity and speed, but also to the line being designed to accommodate passengers. By 1836, more than

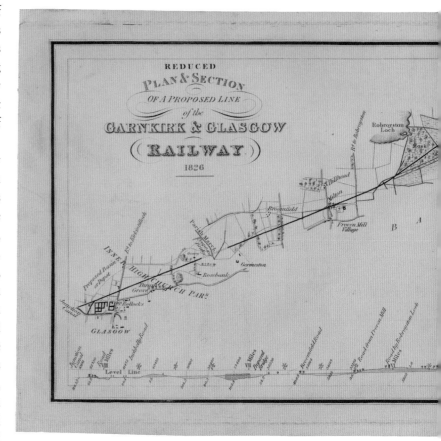

145,000 travellers and 137,827 tons of material were being transported each year. Competition and friction between the local railways eventually resulted in the Garnkirk and Glasgow linking directly with the Wishaw and Coltness line and renaming itself to the more appropriate Glasgow, Garnkirk and Coatbridge Railway in 1844. Subsequently, it amalgamated with the Caledonian Railway. Despite a proposal by Grainger and Miller in 1829 to link the railway with the Clyde, opposi-

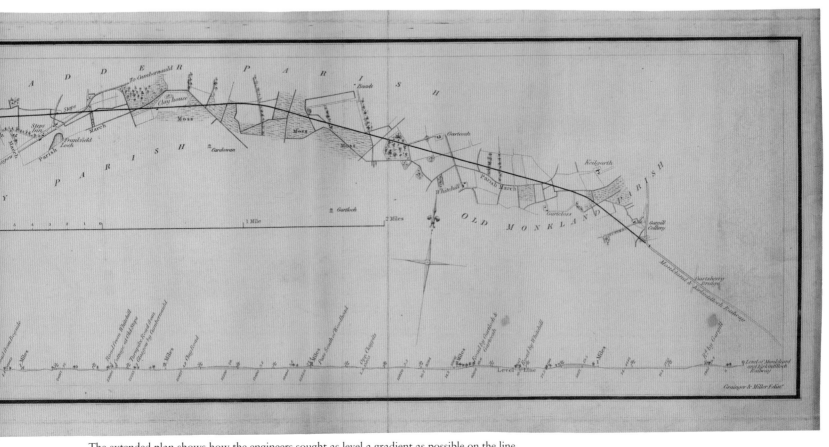

The extended plan shows how the engineers sought as level a gradient as possible on the line

tion from the Blythswood estate trustees and the proprietors of the Forth and Clyde Canal prevented any such development.

The opening of this line marked a change in the working relationship of the two engineers and subsequently they were engaged on separate projects within the partnership, which lasted until 1847. Miller continued as the engineer to the Glasgow, Paisley, Kilmarnock and Ayr Railway, built between 1837 and 1843. Even more importantly, he was engineer for the Edinburgh and Glasgow Railway, which included the 36-arch Almond Viaduct. On retiring from this career, he sat in parliament as Liberal M.P. for the city of Edinburgh from 1868 to 1874. Grainger served as a member of the Edinburgh City Council and was, like Miller, a Fellow of the Royal Society of Edinburgh and President of the Royal Scottish Society of Arts but died in 1852, as a result of injuries he sustained in a railway accident near Stockton-on-Tees.

1827

Recording policemen's beats: the 'sketch of the stations of the several watchmen'

This is a small plan but it has much behind it. The growth of what generally can be gathered under the term 'city administration' has already been seen in the indication of parish and royalty boundaries on earlier maps of the city. With a burgeoning population, there was an increased need to protect and safeguard both properties and citizens. Glasgow at this time was growing faster than any other equivalent city in Western Europe. Between 1801 and 1831, the population rose from 77,385 to 202,426. Traditionally, Scottish burghs employed either hired watchmen or volunteers as part of a guard or city watch. Glasgow was to set an early example in 1779 by appointing an inspector of police and establishing a body of eight officers. Both this and a further attempt to raise such a force nine years later failed, due to objections from the citizens to paying rates for its upkeep. However, on the 30th June 1800, the Glasgow Police Act received the Royal Assent, thereby establishing such a force but, possibly of greater importance, guaranteeing a stable financial resolution.

It is a reflection of the involvement of the Council in the development of services that this act preceded the creation of the London Metropolitan force by 29 years. By the autumn of 1800, John Stenhouse had become the city's first Master of Police and he commenced enlisting a body which soon grew to two sergeants, six officers and sixty-eight watchmen. At first, it was based in Candleriggs but, after a couple of moves, a purpose-built Central Police Office was established in South Albion Street in 1825. This is clearly depicted on the map. That year, John Graham, a city wine merchant, took over as Superintendent of Police and held the office until his death in 1832. Graham improved the force considerably and put it on a more professional footing. One of his policies was to recruit only men under 45. He was appointed marshal for all public parades subsequently and it may be a reflection of how highly his services were valued within the city that the Council agreed

John Graham, *Sketch of the Stations of the Several Watchmen within the Royalty of the City of Glasgow* (1827)

to pay half his funeral expenses.

Glasgow's magistrates had displayed an innovative approach in their thinking behind the intentions to improve public safety. Elected commissioners would be responsible for the administration of the force and the officers would wear a recognisable uniform with a badge. Lanterns, staves and rattles were to be carried. Beat patrols were established to supplement the static duties of the watchmen as an early mode of 'preventive' action. Many watchmen were elderly and initially their duties included lighting and sweeping the streets. This was halted in 1804 but they continued to call the hours at night from their sentry boxes.

The *sketch of the stations of the several watchmen* accompanies a pamphlet issued by the Commissioners in 1827 which lists the 102 watches and 24 wards covering the city. These are all delineated on the plan and given a separate colour wash, while the wards are identified by Roman numerals. Each beat and its range are specified in detail in the text – for example, 'No.62. George Street, east from Portland Street, Bun's Wynd and Shuttle Street, north of College Street'. In consequence, most streets are named on the plan but, as separate burghs set up their own forces, the depiction covers the royalty alone. In general, attention was paid to principal streets, with patrols often not extending to the poorer districts, where there were fewer ratepayers. Features named on the plan include the jail and public offices on Saltmarket, the infantry barracks off the Gallowgate, the Magdalene Asylum and Tennent's Brewery. Whether this last mentioned building was a source of trouble or inspiration to the force is a matter for personal conjecture. That the New Bridewell prison off Duke Street is not shown possibly underlines the difference between policing and the treatment of offenders.

At about the same time as this pamphlet appeared, the first of a sequence of regulations and instructions was issued giving the watchmen a clear list of duties. These included taking notice of any idle or suspicious persons, keeping an eye on disorderly houses and ensuring that lamps were properly lit, sewer gratings cleared and carts removed from the streets. This was a time of considerable social change and unrest, as well as political

ABOVE. Anon. Watercolour of a Glasgow policeman, from an album of old uniforms of the Glasgow Police Force (nineteenth century)
OPPOSITE. The plan identifies all of the beats of the city police force within the Royalty and, hence, there is no coverage of the area south of the river, which was a separate barony

radicalism. Vagrancy was possibly the most serious concern confronting the police. Declining wage levels and the volatility of the local economy led to a large number of unemployed or poor migrants frequenting the city streets. Begging was increasingly discouraged, reflecting not only a growing desire to better regulate and improve city life but also a hardening of attitudes

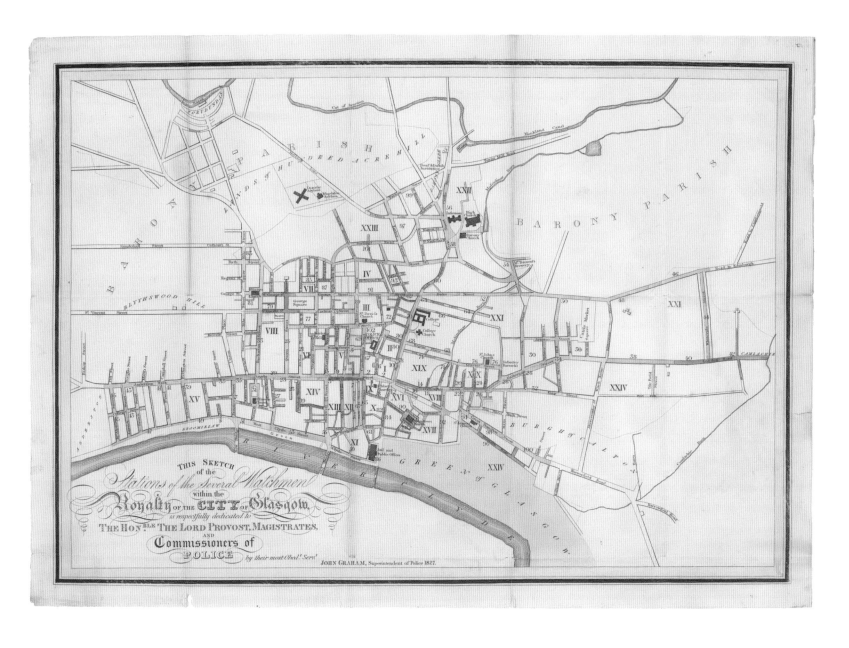

towards the urban poor. While police duties were to expand considerably, large scale civil disturbances or riots could not be handled by the force alone and required the support of the military, local yeomanry, volunteer forces or special constables. This was brought into sharp focus during the unrest of the Scottish Insurrection of 1820, and the 1821 Glasgow Police Act introduced many of the reforms alluded to in this plan.

A similar experience occurred in the 1919 George Square riot, giving Clydeside a further reputation for radicalism. The city was to retain a separate force until 1975 when it amalgamated with other local forces to create Strathclyde Police, now part of the national body, Police Scotland.

1828

'On a scale peculiarly well adapted for Counting-houses'

Multi-sheet maps of any city are expensive and rather unwieldy documents. Both John McArthur and Peter Fleming had created reduced plans from a single plate relatively soon after publishing their own large-scale surveys. In this present example, David Smith waited seven years after his own six-sheet plan before preparing a more compact version. Part of this delay may have been financial. As early as December 1825, the Glasgow booksellers and publishers Walter Wardlaw and John Cunninghame had advertised a new city plan based on actual survey and indicating recent improvements, particularly in the west. As with Smith's earlier work, Robert Scott was to be the engraver and subscriptions at the sum of one guinea apiece were called for.

It is possible that this price was set too high, for nothing more was mentioned of the project for nearly two years. Certainly, an engraved plate was ready by late 1827, to be dedicated to Archibald Campbell, M.P. for the city between 1820 and 1831. By this time, land on his estate was being feued out for the construction of a number of fashionable residences designed for Glasgow's wealthier citizens seeking to quit the inconveniences of the older parts of the city. Frequently, they headed to what Cleland was subsequently to describe as 'the splendid town of Blythswood'. The later press announcement has two interesting features. It reflects a noticeable price reduction, where the plan was now offered as suitable for businesses at 12/- plain or 15/- coloured, but also requests the public to offer 'any hints or suggestions' on early impressions. These may have held back the map's publication until the following year. Consideration of the existing 1827 'proof' copy shows marked differences from the finished work.

Regardless of these delays, this is a most detailed and well engraved plan, encompassing the growing city in a single sheet. The talents of both Smith and Scott are evident in the level of information they managed to include but the several extru-

David Smith, *Plan of the City of Glasgow and its Environs with all the Latest Improvements* (1828)

sions, where the features indicated burst through the map's border, suggest that even this could not contain the city's expanding form without difficulty. On all four sides, the depiction literally breaks out of the confines of the frame to underline the impression of the energetic growth of Glasgow. In some areas, such as at Gallowgate Toll, whole sections extend beyond the margin to identify roads, works and houses.

Within the map itself, there are further signs of dynamic change, with the addition of several suggested new roads indicated by pecked lines sweeping across original features. It is in the identification of what can be described generally as transport developments that the map most reflects new elements in the city. Smith records the arrival of the Paisley and Ardrossan Canal at Port Eglinton and the Garnkirk rail link to the coal basin at Barrowfield, as well as the renaming

of a few streets and the re-positioning of some toll bars. A new five-span bridge shown crossing the Clyde immediately downriver from the Hutchesontown wooden footbridge seems to pre-date its construction since building only began in 1829. Ferries are now shown crossing the river beside the new and old harbours and the Lancefield Basin is identified.

Possibly of greater significance is the extended coverage west of the city centre. The gradual development of Blythswood Hill, in what has also been hailed as the second new town, during the intervening years since Smith's earlier map is noticeably striking. This was the first area in the expanding city where cartography met topography. Up until this point, the grid pattern had been laid out across the relatively flat lands of Meadowflats and Ramshorn, as well as south of the river. The Blythswood estate, however, covered the more undulating drumlin landscape which stretched from Argyle Street north over Garnethill to Cowcaddens. As with his earlier survey, Smith had problems in attempting to show both the new layout and the corrugated nature of the ground. In truth, his is at best a poor compromise and the immediate geography is only hinted at by the hachuring he employs. Combined with an odd configuration of the street plan south of Blythswood Terrace, the identification of proposed new roads and the retention of unexplained engraver's lines, this is one area where the cartography is both confusing and unsatisfactory.

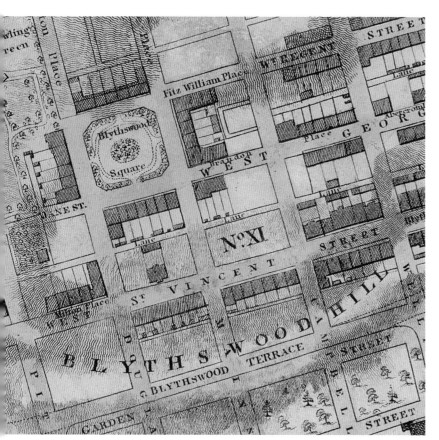

Many more streets and properties have been added to what is now a more definite grid pattern and, with an eye to potential sales, a considerable number of proprietors are named. Surprisingly, despite this extended coverage, the map has no dedication to Campbell and the plain box of the title is only part of a general reduction in ornament. This also extends to the removal of the inset map of 1779. To aid clarity, some map information, such as district names, has been removed to compensate for the smaller scale but this so serves to reflect Scott's ability as an engraver that the only table of references is to the parishes, which are denoted by Roman numerals. Other additional features include the observatory on Hill Street, the Botanic Garden, Knox's monument and the

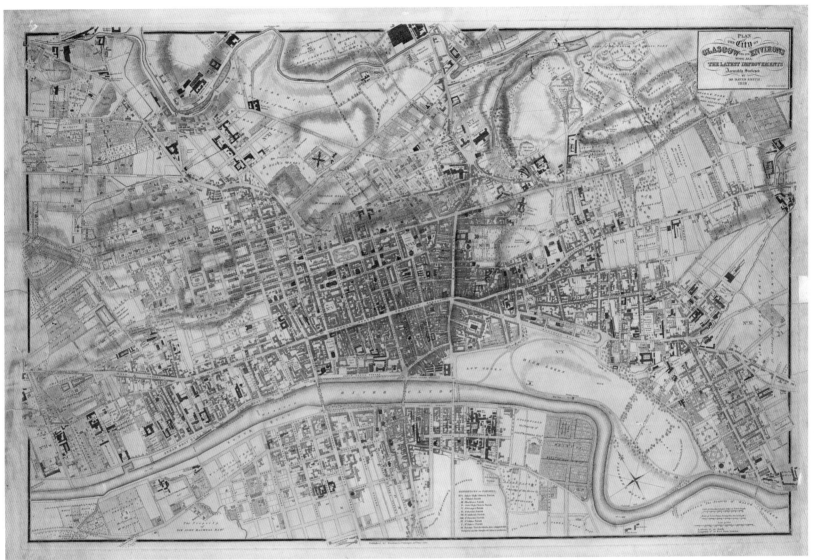

OPPOSITE. The complexities of representing a tentative street development on top of a more varied topography was a particular challenge for the engraver

ABOVE. The way in which the city breaks through the margins of the map emphasises a sense of dynamic growth in the urban pattern

New Gorbals burial ground. Several more industrial units are marked east of Clyde Street but, other than the Blythswood layout, Smith has added surprisingly little from the intervening period. This is particularly true south of the river.

Nonetheless, the publishers may have been well content with the end result, for a later press notice indicated a smaller-scale reduction planned in 1830. Smith's career was to last nearly fifty years but, like several others in his line of work, he turned his attention increasingly to civil engineering contracts.

1832

Parliamentary reform and the extension of the city

This plan is to be found in the *Reports upon the Boundaries of the Several Cities, Burghs, and Towns in Scotland, in respect to the election of members to serve in parliament* as part of the legislative process which became known as the Great Reform Act of 1832. While the more famous legislation, the Representation of the People Act, introduced considerable alterations to the electoral system in England and Wales, a separate act for Scotland mirrored the changes north of the border. The notorious 'rotten' boroughs did not really exist in Scotland but the new act resulted in an increase in the number of Scottish MPs from 45 to 53. Both Glasgow and Edinburgh were now able to return two members to parliament but, significantly, the number of electors rose from under 5,000 to over 65,000 and for the first time included householders with property valued at £10 or more within burghs. It was a start at widening the franchise or, in other words,

improving the voting system, but even after this date only one in eight of the adult male population could vote. The significant shift in power was from the landed aristocracy to the growing urban middle classes. It would be more than another 85 years before women got the vote.

In preparation for the implementation of the Act, each burgh in Scotland was visited between November 1831 and February 1832 by two commissioners, William Murray and Captain J.W. Pringle of the Royal Engineers, to establish new boundaries, both by written description and plan. Clear instructions were given to the commissioners to ensure that the revised boundaries would be precise and unambiguous and the 73 separate plans which accompany the *Reports* are designed to illustrate definitively their delineation. Frequently, these identify permanent urban features to aid the illustration. Burghs were visited at least twice and, given the dates, it

[James Gardner], *Plan of Glasgow*, from *Reports upon the Boundaries of the Several Cities, Burghs, and Towns in Scotland, in respect to the election of members to serve in Parliament* (1832)

GLASGOW

ABOVE. The proposed boundary of the city used many permanent features, including the River Kelvin, and provided an extended area for city growth in the east

OPPOSITE. Gardner depicted only a limited number of individual buildings but here identified the proposed Parliamentary Road

suggests that this investigation was carried out within a very tight time schedule. The extensive burgh reports included information on their contemporary trade and industry, as well as details on population, the number of houses valued above the £10 figure, assessed taxes payable and the general circumstances of the town. By 1831, Glasgow's population was in excess of 202,000 and had more than 6,600 houses in the £10 and over category.

The reports also contained details of the proposed boundaries, along with the certain fixed points which appear on the

plans. In Glasgow's case, the boundary began 'on the west side of the town, at which the River Kelvin joins the River Clyde, up the Kelvin river to a point (2) which is distant 150 yards (measured along the River Kelvin) above the point at which the same is met by the park wall which comes down thereto from Woodside Road'. Murray and Pringle used the courses of rivers and canals, where possible, and connected these by straight lines to those selected points or roads considered permanent. In their report, they noted that, while the march would include most of Glasgow and its suburbs, Partick and certain industrial concerns (such as the Port Dundas distillery and the St Rollox foundry) had been left out, largely to ensure the integrity of the boundary line. Significantly, a large portion of open ground was included within the new city area, being deemed necessary 'to so rapidly an increasing town'. This is particularly true north of the Clyde, where the lands of Golf Hill, Whitehill and Haghill appear remote from the built-up area.

Unlike other contemporary town plans, notably those by John Wood, these were all drawn in a similar fashion and at the same scale of six inches to a mile, despite being executed by a number of London engravers, namely James Gardner, Henry Martin, Josiah Henshall, Thomas Ellis and Benjamin Davies. This uniformity contrasts with those plans prepared for England and Wales, which were produced at a range of smaller scales. Although the Glasgow plan has no attribution, the likelihood is that it is the work of James Gardner, who was responsible for the neighbouring maps of Edinburgh and Leith, Aberdeen, Paisley and Dundee. Based in Regent Street, Gardner was additionally a publisher, map and globe-seller and sole agent for the sale of Ordnance Survey maps. He was a founding

Fellow of the Royal Geographical Society in 1830 and engraved maps for the *Transactions of the Geological Society*.

Inevitably, the plan gives only a basic street pattern for the city centre, with general detail in shaded block format rather than the careful delineation of all the key buildings and roads. Here again, purpose affects what is shown. There has been no attempt made to indicate the closes and wynds which ran off the High Street and Saltmarket but major thoroughfares have been identified and several of the most important public buildings named. There is no intention to display any new developments and the layout of both the Gorbals and Sandyford-Woodside districts tends to show only isolated buildings. Quarries, bridges, tolls and country houses in the immediate neighbourhood are identified. The only colour used on the plans is the proposed boundary (in red) and water (in blue). Certain mistakes in the naming of features, such as 'Patrick Bridge' and 'Cannig Street', suggest a rushed preparation for the finished product. This adds weight to the belief that the plans may have been based on earlier depictions of both the burghs and local estates, particularly likely as some of the neighbourhood information is unique.

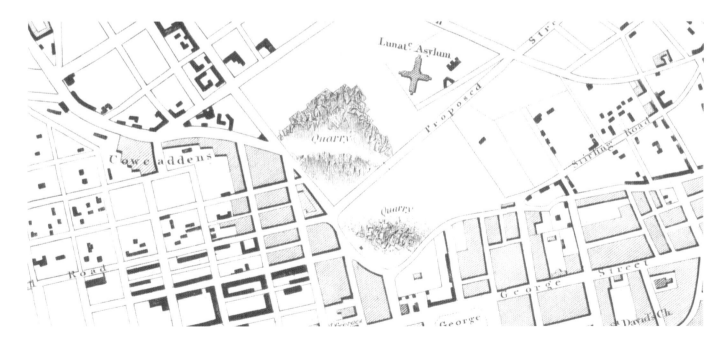

1835

The City in black and white

Apart from a couple of related lithographic plans produced in 1830 by Berkeley King and based on Smith's map of 1828, these two depictions mark a major change in the way the city was represented cartographically. King was a landscape artist, engraver and lithographer who worked in London and specialised in moonlit scenes up until about 1835. He subsequently produced maps for the Quartermaster General's Office.

Prior to this date, all the maps covering Glasgow were prepared either in Scotland or for a particular purpose, specifically parliamentary reform. With the appearance of Dower's plan, the advent of commercially produced representations of the city for a wider audience was heralded. John Crane Dower (1791–1847) established his family's map engraving and printing business in Pentonville, London in 1820. He is recognised as an engraver of an extensive range of railway, country, county and geological maps, as well as a general world atlas.

In particular, he prepared maps for both Thomas Moule's *The English Counties Delineated* (1830–35) and Christopher Greenwood's *Atlas of the Counties of England and Wales* (1834). He also engraved the county maps, surveyed in part by William Fowler, of Fife and Kinross, Berwick, Midlothian and East Lothian. Glasgow appears to be his only Scottish burgh plan.

This depiction of Glasgow was published by William Orr, a close associate of the Edinburgh publishers Robert and William Chambers. With a location in the heart of London's publishing area, Orr himself was a publisher's agent with a flair for distribution. He was involved in the general commercial side of the business, specialising in natural history, maps and periodicals, especially Charles Partington's *British Cyclopaedia*, produced in the 1830s and issued in ten volumes. It now seems most likely that this Glasgow plan first appeared in 1836 in the relevant volume of the *Cyclopaedia* relating to

John Dower, *Glasgow*, from Charles Partington, *British Cyclopaedia* (1835)

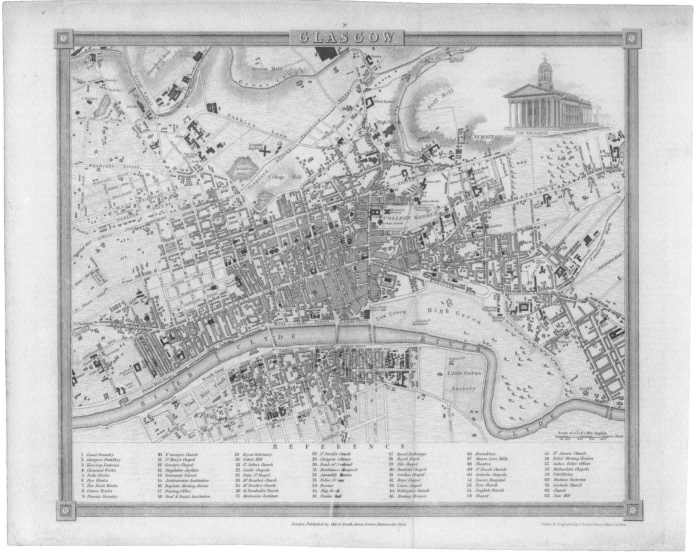

ABOVE. Dower's illustration of the Royal Exchange and the introduction of a table of numbered references were novel features of this plan

OPPOSITE. The adaptation of Dower's plan by Archer in this white line engraving

literature, history, geography, law and politics. Certainly, Dower engraved several other maps and plans for the series.

Dower based his engraving mainly on Hugh Wilson's 1830 plan of Glasgow, which was itself reliant on Smith's depiction of two years earlier. Although a slightly smaller area

than Wilson's is covered, the plan provides a good and clear overall image of the city, following the original source in its attention to the identification of works, factories and foundries. While many previous maps were often illustrated either by decorative title or dedication cartouches or by insets

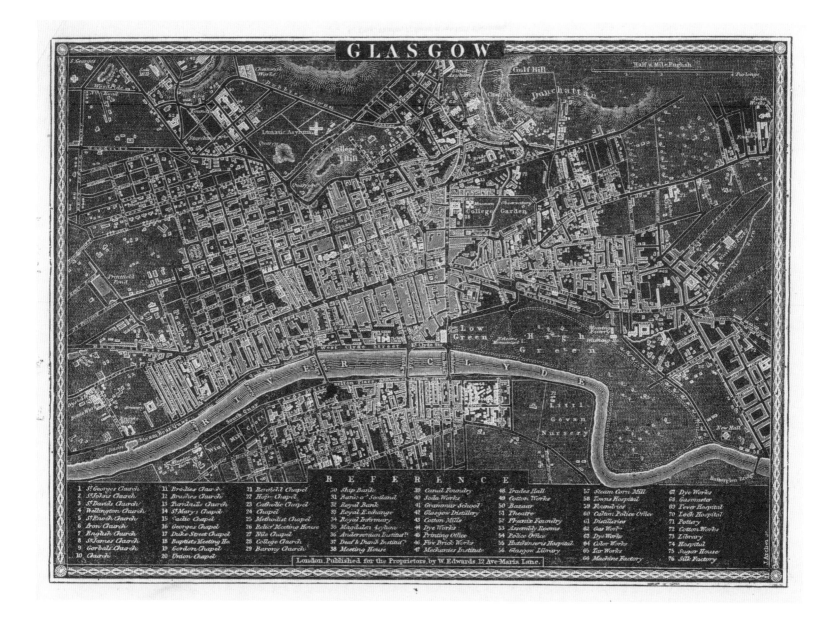

of older plans, this Dower example introduces another innovation with the placing of an illustration, in the north-east corner, of the New (Royal) Exchange, converted from William Cunningham's mansion by David Hamilton in 1827–32. This is added to by a further Dower alteration in the table of references below the map. While many numbers refer to individual

churches or public buildings, the engraver has used others to identify particular types of industry (e.g. 4: Chemical works). In this way, Dower's version is able to give an immediate impression of certain district specialisations within the city – for example, the concentrations of foundries on Washington Street and dye-works in Hutchesontown. This cannot be said

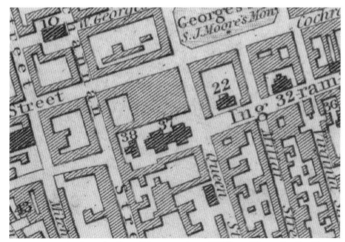 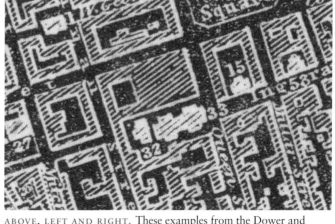

OPPOSITE. Joshua Archer, *Glasgow*, from William Pinnock, *Guide to Knowledge* (1835)

ABOVE, LEFT AND RIGHT. These examples from the Dower and Archer plans give a good impression of the limitations of the white line technique, with a notably poorer quality of definition and a reduction in detail

to be universal for the map does not give any clear sense of the clustering of cotton works in the Calton-Bridgeton area. Unlike its source, Dower has re-introduced hachuring and this is particularly effective in identifying the city's quarries, Golf Hill and College Hill.

This depiction was to be adapted in a unique way by the engraver Joshua Archer as an illustration in William Pinnock's *Guide to Knowledge*. Pinnock (1782–1843) was in the vanguard of popular education and his *Guide* appeared in weekly penny issues of eight pages, as part of a general trend in publishing cheap instructional works for a wider market. He specialised in this type of literature and catechisms, also producing a successful series of county histories. His *Guide* was regularly illustrated by town plans, celestial and county maps. The Glasgow plan appears in the issue for 30th April 1835, which was dedicated to a detailed description of the city and its present state.

It is distinctive in being an example of white line engraving, in which an image was cut or punched into the surface of a wooden or metal block. When printed, it appears white on the inked background. This was not a new technique and, during the nineteenth century, it gained a limited popularity. However, it was difficult to guarantee a good print because the technique frequently resulted in inevitable variations in the inking of a wide surface area unbroken by any engraved marks. There were limitations in the amount of detail which could be depicted on the image and the results often appeared crude. It was certainly not an ideal method to display the intricacies of a town plan and was not developed further.

Archer was a draughtsman and prolific engraver who had produced work for Thomas Dugdale's *Curiosities of Great Britain*, but in the year of this plan he was declared bankrupt. He was responsible for the engraving of several of the county maps and a plan of London which appeared in Pinnock's *Guide* and is recorded as working in Pentonville, thus strengthening the connection with John Dower. A comparison of the maps shows up changes in the placing of names and the indication of features. Despite the difference in style, the identification of quarries and hills is remarkably clear. One obvious alteration is the table of references, which has been redrawn completely and extended to reduce the lettering on the map itself. This is not a well known map but, along with its source, it literally does display the city in black and white.

1836

A corporate production: Knox's map of the Clyde

One of the striking features of this map of the Clyde basin is the number of individuals who were involved in its production. Tellingly, their businesses were all located in Edinburgh. James Knox is a relatively little known Scots cartographer but he was clearly a talented practitioner. Based in the capital, his name is linked initially to surveys around the city dating from 1802, and he was trained by John Ainslie. His first published work was a county map of Midlothian, which appeared in 1821 and was included in John Thomson's *Atlas of Scotland*. Like Thomson's own work, this map seems to have caused the surveyor financial problems for, around the period 1822–26, Knox had to sell off maps and plates to pay his creditors. Subsequently, he was responsible for plans of Edinburgh and Paisley but also drew maps of the Forth (1828) and Tay basins (1831).

This Clyde sheet has been drawn in the same style as his other river basin maps but in each case a different engraver was employed. In this instance, the map records the participation of two separate figures. While the depiction of hills and parklands was the work of William Home Lizars, the outline and lettering was left to the lesser known John Muir. Lizars (1788–1859) was based in St James's Square and was closely associated with the work of Sir Walter Scott. His map engravings cover the period from the end of the Napoleonic Wars to the middle of the century. In 1820, he had prepared a small-scale map of the Clyde and the Western Highlands for James Lumsden's *Steam Boat Companion* but this is an entirely different representation. The finished work was issued by two of the capital's publishers. John Anderson junior was a second-generation bookseller based on Edinburgh's North Bridge, while William Hunter, who also ran a circulating library, had his premises in the New Town, in South Hanover Street.

Not only is there an absence of local involvement in production but there also seems to be no mention of the map

James Knox, *Map of the Basin of the Clyde* (1836)

ABOVE. While the use of colour is limited to the depiction of parish boundaries, the employment of two engravers and, particularly, William Lizars resulted in a map with high levels of topographic detail

OPPOSITE. Knox's intention to publish an accompanying textual account of the Clyde Basin led to several items of historical interest being marked on the map, as seen here off Gourock

in the Glasgow newspapers. In fact, the only announcement of the work which has been traced appears in the publisher's own *Tourist's Guide through Scotland* of 1837, among a list of other maps and guides and described as 'preparing for publication'. Interestingly, the catalogue also records an accompanying text by Knox, *The Topography of the Basin of the Firth of Clyde*, which was to cover the same districts as the map, with an account of their agriculture, commerce, scenery and antiquities. This does not appear to have been published. Nonetheless, this map is a detailed depiction of the river and firth covering an area between Arran and Carstairs, and from Buchlyvie to Cumnock, thereby illustrating most of the counties which are contiguous to the city.

This is a map which deserves close scrutiny and repays careful consideration. Landward areas are shown with a notable degree of detail, particularly in the Clyde Valley, where the initial impression is of a confusing jumble of names. It is packed with a range of topographical information, however, and is much clearer in areas away from the river. The engravers have employed a sophisticated range of lettering to distinguish various features but there is often a lack of distinction between houses and industries – for example, the Clyde ironworks are somewhat lost among the host of other properties. While the names of several owners appear beside their country seats, these tend to be where there is sufficient space for their inclusion. Many additional notes provide details of historical interest (e.g. at Gourock Bay, 'Here the Comet, Steamer was run down, 1825'; to the north of Ayr, 'Ruins of hospital for lepers endowed by King R. Bruce') and hint at the antiquities Knox intended to cover in his accompanying topography. This is strengthened by the delineation of the Antonine Wall and

the identification of the Roman fort at Camelon, combined with the appearance of certain Latin names (e.g. 'Clota Aest').

Sandbanks and offshore soundings are indicated in the estuary and the line of ferry routes from Glasgow to selected destinations suggests the main channel up to the city. Drainage is well defined and much care has gone into the marking of tributaries. This is continued into the depiction of upland areas, by hachuring, and mosses, by stippling, where the change of engraver has produced a more effective image in this instance. In general, the map provides a good impression of the road network, again emphasising the concentration round Glasgow as a centre for the whole region. Features such as the Garnkirk Railway, the Paisley Canal, tolls, quarries and mines also find their place.

Taken overall, it is really up to the reader to decide whether or not the intensity of detail is acceptable or irritating. Did Knox try to do too much on the one map? He did succeed in publishing a companion topographical description to the Tay basin which appeared in the same year as the associated map (1831). Unlike his cartographic coverage of the other firths, this Clyde map was re-issued in 1837 and 1838, which may suggest that it was, indeed, a successfully popular product.

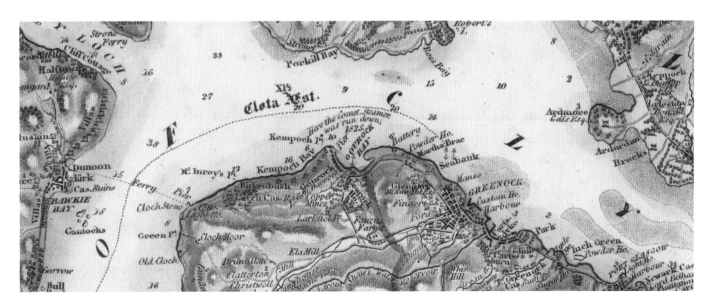

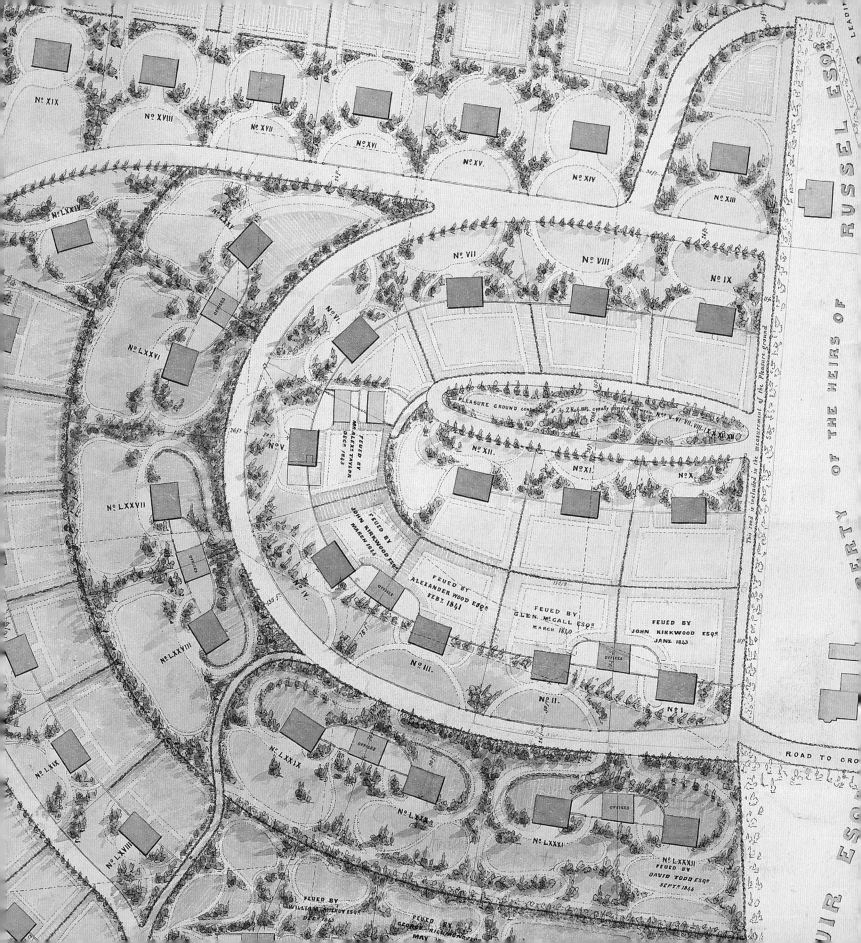

1840 & 1849

The best laid schemes: estate plans north and south of the Clyde

These two plans illustrate the variety among the different proposals to develop estates at the city's fringes on either side of the Clyde. They are also excellent examples of the contrast between initial intention and eventual outcome. Both were designed as exclusive suburbs for a middle class that was moving away from the increasingly overcrowded and unpleasant conditions of the city centre. The proposals also underline the changes being experienced in the profession, whereby a growing number of plans were being produced as much by architects and civil engineers as by surveyors themselves.

During the early nineteenth century, Partick grew from a

Alexander Taylor, *Feuing Plan of the Lands of Partick-Hill* (1840)

village of about 1,000 people to over 3,000 by 1840. Further change followed the opening of the Tod and MacGregor shipbuilding yard at Meadowside in 1844 and the subsequent increase in local heavy industry. Unlike other developments, Partickhill's relatively isolated location did not deter speculation and, despite its remoteness from the centre of Glasgow, its proximity to Partick resulted in it becoming one of the earliest designed estates in the West End. In 1840, Alexander Taylor prepared this feuing plan of 98 plots for the owner, William Hamilton. Taylor had begun to practice as an architect and civil engineer in Edinburgh, before moving west. Based in St Vincent Street, he was responsible for several other feu proposals in the city, including Ibrox, Govan and South Park. Possibly his best known Glasgow commission is Royal Crescent.

This plan is an interesting contrast between regularity and a more fluid design. Inspired by the landscape, Taylor made use of the features of the drumlin at the crest of the hill in proposing four concentric crescents curving round the upper slopes. A mix of spacious villas, semi-detached houses and short terraces was intended. An outer loop would have dramatic views over the surrounding lands and link to the layout to the south, where a more rigid pattern of villas facing a central square would be separated from Dumbarton Road by a row of terraces and gardens.

As a working document, the plan shows that the best parcels on the summit were early occupied, with eleven plots feued by December 1845, including one to Taylor himself. Within twenty years, however, the original concept had begun to change. The division into smaller units resulted in an increased density. Terraces were replaced by tenements, particularly when the Stobcross branch of the North British Railway created a barrier to the west. Today, the crown of Partickhill still survives as a quiet enclave where the lack of through roads allows it to retain an air of exclusivity. It stands in marked contrast to the regular pattern of tenements which is the

hallmark of the nearby localities of Hillhead, Hyndland and Dowanhill. The central square of the original plan still exists as the West of Scotland Cricket Club's ground. Known as Hamilton Crescent, it was the location for the first international football match between Scotland and England in November 1872. Taylor died in 1846 before seeing too many of these alterations.

South of the Clyde, in 1849 David Rhind (1808–1883), an Edinburgh-based architect, was commissioned by Sir John Maxwell of Pollok to design the layout of the Pollokshields area on farmland adjacent to the old Shields steading. Maxwell had earlier employed Peter MacQuisten, a local surveyor, to draw up a plan for the lands of Kinning House and his 1834 proposal for a 'model new town' shows a design centred on a large circus. Rhind was to introduce a similar feature in the western section of his plan (Bruce Circus) but this was never realised. As an architect, he was employed chiefly in design proposals for the Commercial Bank but was also responsible for the plinth for the Scott monument in George Square and Edinburgh's Daniel Stewart's Hospital. In 1855, he was elected President of the Scottish Society of Arts. An earlier feuing plan of Merchiston, a prosperous residential suburb in Edinburgh, dated 1844, suggests that Rhind was comfortable with designs which used space generously.

In this proposal for Pollokshields, he similarly utilised the advantages of the estate, with broad winding roads and an extensive plot size – here, frequently more than half a hectare. While it was later criticised for being unremarkable, the plan displays a blend of two distinct units. East Pollokshields has been laid out on a north-south axis centred on an extensive area of gardens faced by a combination of terraces and crescents while, to the west, the plan indicates a less regular pattern of villas. The eastern terraces acted an additional screen blocking out the view, particularly of the Barrhead branch line and industrial areas which lay further east, from the higher quality residences of the western villas.

OPPOSITE. Taylor's plan made use of the immediate topography to provide a blend of curved terraces to the north and a more regular frontage on to Dumbarton Road

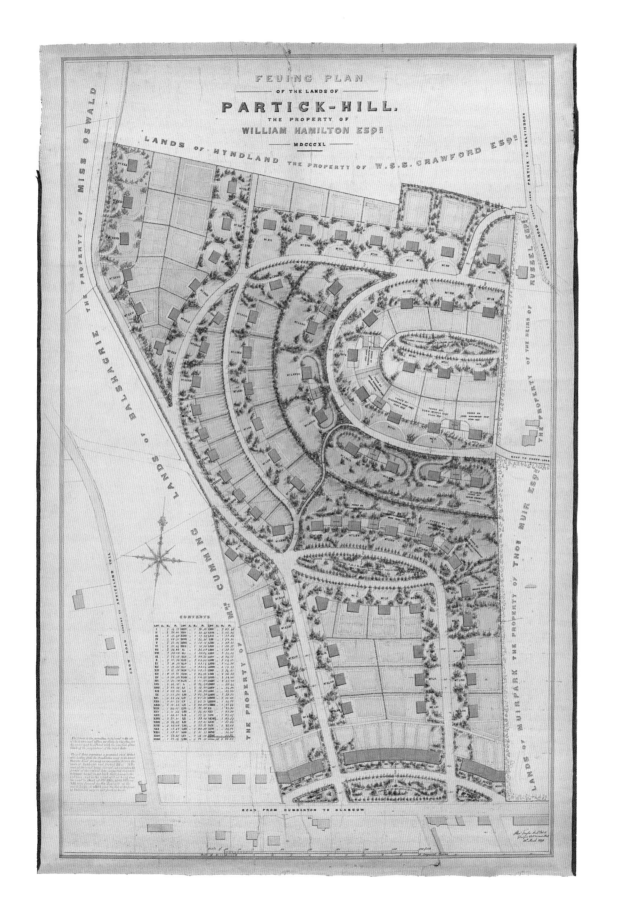

Like Partickhill, the subsequent development of the area did not 'go to plan'. The eastern portion of the proposal was changed quite considerably by the introduction of a greater number of tenements, increasing the housing density, and the intrusion of shops. In addition, the alteration of the central gardens to a small square reduced the sense of space in the locality. However, strict feuing conditions controlled by the factor, William Colledge, meant that West Pollokshields retained its garden suburb layout and it remains relatively intact, consistent with the amenity of the neighbourhood. A degree of exclusivity resulted from the limited access to the area, bounded as it was to the north and east by railway lines and to the south and west by the lands of the Pollok estates. By the end of the century, many substantial houses, designed in a variety of styles, had been constructed. It is now one of Glasgow's designated conservation areas. As with the evolution of the West End suburbs, Pollokshields was only one of a number of schemes proposed for the south side. Elsewhere, Langside, Strathbungo, Shawlands and Govanhill would be laid out in the ensuing decades.

David Rhind, *Feuing Plan of the Lands of Pollok Shields* (1849)

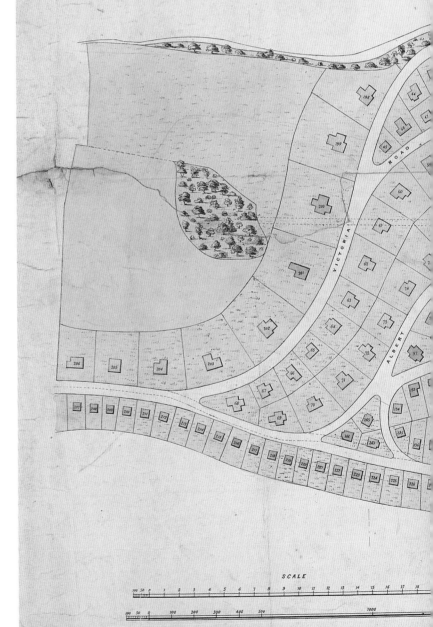

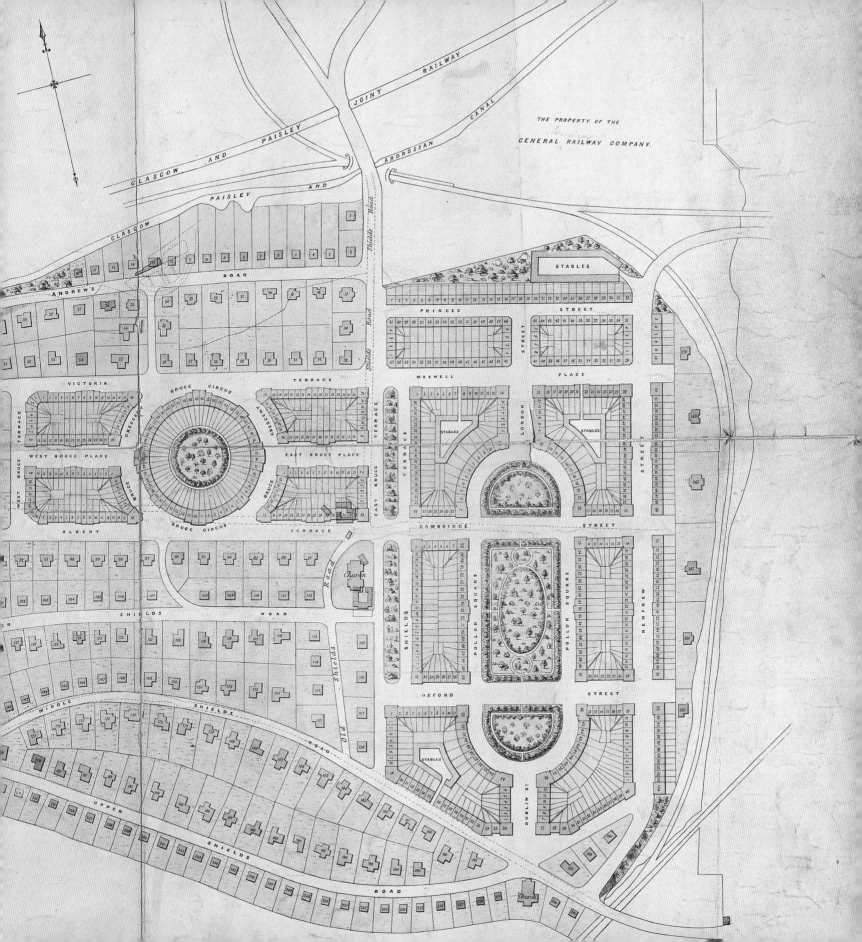

1841

Lithography used to illustrate for an extended audience

In September 1840, the British Association for the Advancement of Science met in Glasgow for the first time. Local participation included reports on various aspects of the statistics of the city, particularly one on its population, trade and commerce by James Cleland, while Sir William Hooker crowned his time at Glasgow by presiding over the botany and zoology section meetings. It is possible that a heightened interest in Glasgow led local publisher William McPhun (1793–1877) to produce a new edition of his *Stranger's Pocket Guide,* which had been published first in 1833. This was accompanied by an updated and more extensive map than those which had been engraved by Joseph Swan for earlier editions of this affordable work. These earlier depictions had been based on Robert Scott's plate dating from 1821, but widened and adapted to reflect the growing city. The printing flaws which were obvious even in 1833 may have been behind a move to a completely new version. Certainly, this map bears

no attribution to Swan and it was to be the last appearance of the *Guide* itself. However, although the *Glasgow Herald* announced its publication in early September, there is no mention of a new map to accompany the work.

McPhun moved his business from the Trongate to Argyle Street in 1840 and the change of address may have been an additional contributing factor for the new edition. In comparison with the previous McPhun map, this improved and less crude image has many features based on a completely different original plate. It seems most likely that the source was the 1839 plan of the city produced by David Smith and James Collie as a revision of Smith's own one-sheet map of 1828. Interestingly, this Smith and Collie depiction was also engraved by Scott. This present McPhun map heralds Glasgow's harbour developments by showing the proposed dock at Windmill Croft and the new City Wharf beside the Low Green but it has also taken elements, such as the railway lines running

---

William McPhun, *Glasgow and Suburbs* (1840)

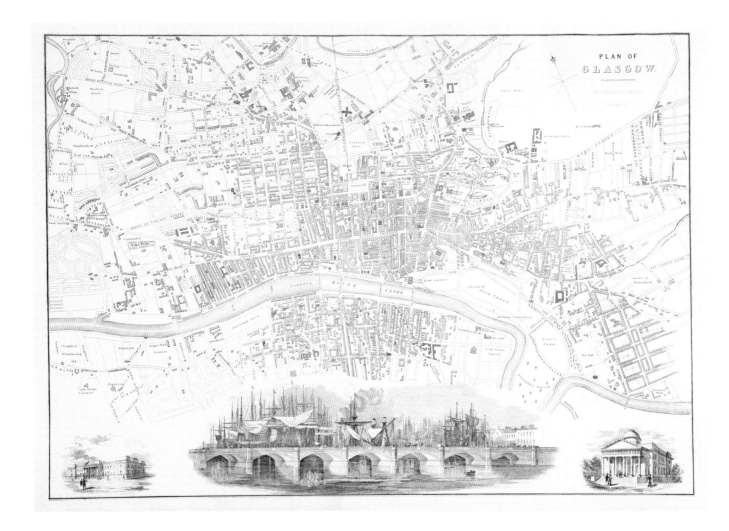

down West Street, from his own earlier plan. Nonetheless, there are significant differences which show this version to be more than a copy, particularly in the indication of proposed street layouts in the Sandyford area, south-west of Kingston and north of Woodside Road. Despite the date, the Royal Botanic Garden is still marked to the west of Clairmont Road, while the Edinburgh and Glasgow Railway is shown coming into its depot to the north of George Square.

The plan has a far wider significance than the merely local. It was to re-appear in three separate states produced for a wider audience by the Edinburgh lithographer, William Nichol, younger brother of James Pringle Nichol, professor of astronomy at Glasgow University. Nichol was based in Hanover Street and it seems most likely that he was related to the Montrose publishers, J. and D. Nichol, for in 1841 the plan was included in their *Glasgow Illustrated in Twenty One Views*. In the same year, it was also included in a portfolio collection of prints as part of *Nichol's Cities and Towns of Scotland Illustrated*. The Glasgow volume was a prototype for the series of views which eventually included Aberdeen, Perth, Dumfries and Montrose. Overall, the works are fine examples of early lithographic printing in Scotland and Nichol

OPPOSITE. William Nichol, *Plan of Glasgow*, from J. & D. Nichol, *Glasgow Illustrated in Twenty One Views* (1841)

ABOVE. The crowded harbour scene below Jamaica Street Bridge

was clearly well placed to produce these, as he wrote the entry for 'Lithography' in the seventh edition of the *Encyclopaedia Britannica*, also published in 1841.

While the adoption of the lithographic process in Scotland was slow initially, it seems to have been well established by 1825. Certainly, this was the date of its first appearance in Glasgow, when James Miller commenced business in Virginia Street. Its early application to map and plan production was appreciated by several local printers, such as Walter Ballantine, and this was to further develop with the introduction of metal plates, rotary steam-powered presses and the transfer technique. By 1840, the Post Office Directory listed twenty-one lithographic printers and one lithographic press maker in the city, with firms such as Allen and Ferguson and Maclure and Macdonald recognised for their involvement in map production. Lithography did not require the engraving of plates and so avoided the problems of plate wear and the

difficulties of regular revision. The process was an effective response to the growing demand for cheap maps and the need to update them in a time of major social change. Apart from the impact of economies of scale, features such as the title cartouche or illustrative vignettes could be positioned as required, making the process suitable for special purposes, such as the indication of street or railway proposals.

This McPhun map is an ideal local example of the process, where the Nichol adaptation of the original introduces three vignettes across the bottom of the plan to portray the Jamaica Street Bridge and the growing harbour, which highlight other developments shown on the map, as well as the Justiciary Building and the Hunterian Museum. This has resulted again in the reduction in the amount of detail of the city south of the Clyde. More importantly, this is an example of a locally produced plan adapted by the addition of new features by other outside publishers for a wider public.

1842

Railways and the westward spread of the city

On 18th February 1842, the railway link between Edinburgh and Glasgow was officially opened. Three days later, services began between Queen Street and Haymarket Stations but, due to the steep incline to Cowlairs, trains were hauled by, first, steam-driven rope, and, subsequently, steel cables up this slope until 1908. The inauguration of this connection between Scotland's two major centres engendered the publication of a series of guides and companions to both the line and the cities themselves. Partly as a result of the increased interest, seven separate maps of Glasgow are recorded for this year alone, including depictions by W. & A.K. Johnston, George Martin, James Mitchell and Thomas Kyle.

No history of Glasgow's cartography would be complete without including something of the work of the Kyle family. As discussed earlier (1807), William Kyle had by far the most significant impact on the city's surveying profession in the early decades of the nineteenth century. By 1795, he had established a school for architectural drawing and surveying in Wilson Street. His associates and pupils included Peter Fleming, David Smith, Robert Park, Andrew Laughlen, Andrew MacFarlane and his nephew, Thomas Kyle. During the 1830s, he worked on feuing plans of the Blythswood estate for the Campbell family. In a professional career extending over forty years, his expertise was recognised by the City Council, who employed him on several commissions, including work on the Clyde navigation, despite retaining John Gardner as surveyor to the burgh. Kyle died in 1837 and was succeeded by Thomas.

William Stuart Stirling Crawford has been considered to be one of several landowners who were less than enthusiastic about the impact of the railways on Glasgow. His estate had largely consisted of arable land and market gardens but by this date it was clear that considerable parts were becoming decidedly industrial in character. In addition, the Port Dundas canal basin had led to a growing intrusion of manufacturing

Thomas Kyle, *Map shewing the Respective Localities of the different Portions of the Entailed Estate of Milton* (1842)

industry onto his Broomhill properties. The plan shows the rail lines running into what would become Buchanan Street and Queen Street stations, as well as the track of the Glasgow and Garnkirk and extended depot beside the city branch of the Forth and Clyde Canal. Projected rail links are also included.

The estate's initial dealings with the railway companies appear inexpert, if not inept. In 1836, the Glasgow and Garnkirk Company were able to purchase six hectares for only £807 and the Edinburgh and Glasgow Company paid a mere £3,000 for the Colston and Broomhill lands. Three years prior to this plan, Kyle had drawn a similar map of Crawford's other Milton properties of Hundredacrehill, Broomhill and Cowcaddens and it is interesting to note that these lands are not identified on this plan. By this date, much of the area

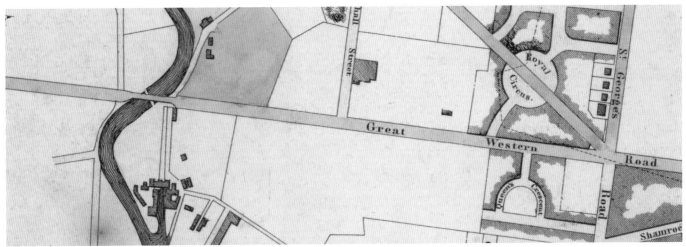

OPPOSITE. Kyle's plan was a working document, showing signs of usage and correction, and was part of series prepared for the Crawford family

ABOVE. Kyle carefully detailed both the Hillhead estate bridge (c.1825) surmounted by the high-level Great Western Road turnpike bridge constructed between 1838 and 1841

around Blythswood, Garnethill and north into Broomhill had already been occupied by housing.

In 1842, a private act of parliament enabled Crawford to apply funds from the estate to lay it out for feuing purposes, as well as to purchase other land parcels to continue development of the area. It is possible that Crawford wanted to keep the railways out and intended to create the right conditions for good quality housing but it may equally be the case that his plans were intended to raise the selling price of his lands. Regardless, the intention to convert land to residential use was part of a general city pattern which can be seen on this plan by the indication of other properties, such as Partickhill and Kelvinside, similarly laid out for this purpose. Kyle's plan clearly relates to the procedure and is valuable in showing how landed estates were being affected by the considerable changes in the urban landscape as a result of transport developments.

While the plan carries a note, dated 15th March 1842, that it was prepared partly from the detailed work of other surveyors, Kyle also states that some of the mapping was based on his own surveys. Certainly, this is an extensive coverage of the city north of the Clyde on the threshold of its advance into the modern age. Although the surveyor's intended purpose was to identify Crawford's land, which has been coloured red, whereas much of the built-up area is only indicated by shaded blocks, the plan shows much more. Despite the somewhat generalised nature of this shading which is markedly indeterminate about where building stops, this is a notably informative, if deceptive, map. Despite being produced as a lithographic print by Maclure and Macdonald, it shows signs of being a working document, for there are pencil additions close to the St Rollox chemical works, as well as the manuscript inclusion of Charles Street and its buildings. In particular, Kyle identifies a far greater number of street names and proposals, notably in some districts west and north of the old asylum, than the impressively detailed depiction by George Martin produced in the same year. Neighbouring properties and their boundaries are located in a broad curve from Whiteinch to Hogganfield. A careful inspection makes it clear that Kyle has drawn on local sources which provide unique information not represented elsewhere. Kyle was to have a further influence on the layout of the west of the city in the 1850s when he worked with the architect Charles Wilson on the layout of Park Circus and Kelvingrove Park.

1842

The last great private survey of the city: Martin's plan

This highly detailed map of the city follows in succession from the plans of McArthur, Fleming and Smith. It is, in effect, the last of the multi-sheet depictions of Glasgow prepared as a separate commission by a private surveyor. However, while this is a large map, it is drawn at a noticeably smaller scale than its predecessors, largely as it has to cover the wider area of the extended parliamentary boundary. It became an important source for later depictions of the city, most notably the directory maps prepared by Joseph Swan, Allan and Ferguson's lithographic plan of 1847 and the Rapkin steel engraving dating from the mid-1850s.

It is the work of George Martin (1809–1875), who was born in Aberdeenshire but seems to have started his professional career about 1834 in the George Street offices of Grainger and Miller, Scotland's principal railway engineers. Certainly, two years later Martin was awarded an honorary medal for a paper describing the Garnkirk and Glasgow

Railway, which he presented to the Society for the Encouragement of the Useful Arts in Scotland. Given that Grainger would become the Society's President in 1849, after it received its royal charter, there were clearly strong links between these men. Soon after this paper, Martin moved to Glasgow, where, in 1837, he set up business as a civil engineer and surveyor. In 1839, he prepared a revised plan of Paisley, originally drawn by James Knox seventeen years earlier, and also applied for the post of resident engineer for the Clyde.

His career was typical of the mid-nineteenth century, where the dividing line between civil engineers, professional surveyors and architects became increasingly less distinct. Much of his surviving work relates to railway proposals and on several occasions his name appears associated with either Grainger or Miller. Examples include the Paisley and Barrhead District Railway in 1845 and the Airdrie and Bathgate line, one year later. As engineer, he worked with Alexander

George Martin, *Map of the city of Glasgow brought down to the present time* (1842)

127

POPULATION IN 1841.

Total Population of Glasgow within the Parliamentary Boundaries 255,650

Total Population of Glasgow and Suburban Districts in 1841
comprised within the same Boundaries as the Population given
in the Census of 1831. 287,154

Total Population of Glasgow and Suburbs in 1831 202,426

Total Increase in 1841 84,728

Shewing that the Population of Glasgow and Suburbs has increased 39.37
Per Cent. since 1831 being a greater increase than that which took place
between 1821 and 1831 by 3.37 Per Cent.

RIVER CLYDE

HARBOUR

BROOMIELAW

Index Map
FOR THE
PARLIAMENTARY
DISTRICTS.

Index Map
FOR THE
MUNICIPAL
DISTRICTS.

To
James Campbell Esq.
The Honorable the Lord Provost
OF THE CITY OF GLASGOW
This Map
IS RESPECTFULLY DEDICATED
By the Publishers
1842.

Index Map
OF THE
POLICE WARDS.

Index Map
OF THE
QUOAD CIVILIA
PARISHES.

Kirkland in the construction of the South Portland Street Suspension Bridge across the Clyde. In the late 1850s, he seems to have had financial difficulties and was declared bankrupt in 1860. However, he subsequently acted as engineering adviser to the burgh of Renfrew and, in about 1870, went into partnership with Robert Dunlop.

Overall, the plan includes an impressive variety of features and pays particular attention to the city's developing transport network. Martin provides a detailed depiction of the growing harbour, centred on the Broomielaw. Soundings for the main channel are marked, while the breadth of the Clyde is given for each ferry and bridge crossing. Intended dock proposals for Windmill Croft and the new harbour at Stobcross are noted, along with the wharf at Great Clyde Street. Despite these new additions, congestion at the quays was so severe that the harbour was almost completely jammed at times, with vessels lying in tiers nine or ten deep. This crowded situation was not properly resolved until the Kingston Dock was built in the 1860s. Unsurprisingly, the newly opened Edinburgh and Glasgow Railway and its station are identified but Martin also marks branch line schemes to the Pollok and Govan and the Hamilton Railways. New street layouts range from the large-scale proposals for laying out Kelvinside and Queen's Park to the delineation of individual roads, such as the proposed Argyle Street continuation. This is very much in keeping with the style of the period, when intended improvements were included as an indication of the currency of the survey. Under-

lining the emphasis on transport and travel, the map is inscribed with a series of concentric circles, drawn at half-mile intervals from Glasgow Cross.

This is the first major plan of Glasgow to be engraved by W. & A.K. Johnston, one of the major Edinburgh publishing houses of the nineteenth century. In the same year as this map was published, they produced a small plan of the city to accompany Willox's *Guide to the Edinburgh and Glasgow Railway*. William and Alexander Johnston began as apprentices to James Kirkwood but by 1826 they were running their own business, eventually acquiring the Lizars firm. By the time of this map, they had been appointed Geographers to Queen Victoria and were based in St Andrew's Square. Subsequently, they were to construct the first globe to depict solely physical features, for which they won several awards at the Great Exhibition of 1851. The company's reputation was later built on a series of significant atlases, including the *Royal Atlas of Modern Geography* in 1861.

It is interesting to compare their engraving style with earlier depictions, particularly in the use of hachuring to indicate both the city's quarries and hill slopes. In general, this is far clearer than, for example, Robert Scott's work for David Smith. On the other hand, a close study shows discrepancies in the continuation of detail, such as building shading and tree symbols, from one sheet to the next. Many recent changes are recorded, including the newly established Royal Botanic Garden on Great Western Road, the Southern Necropolis,

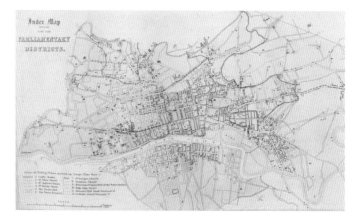

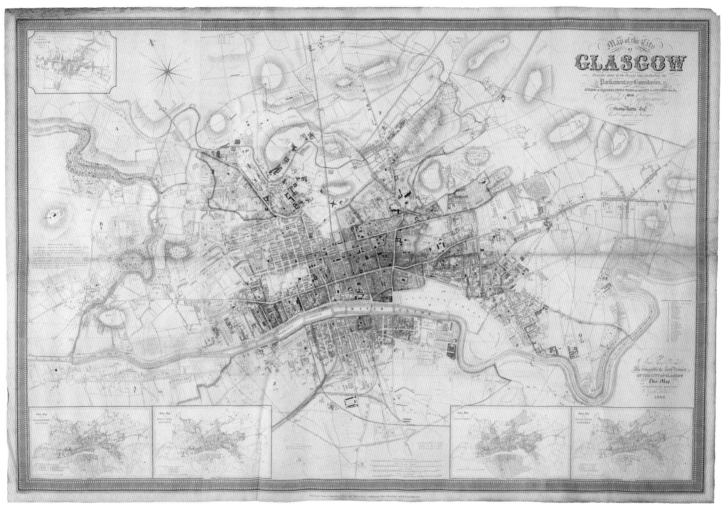

OPPOSITE. The map's appeal to administrators is exemplified by the series of index maps across the lower section, including delineations of the police wards and parliamentary districts

ABOVE. Martin's was the first major plan to show the new parliamentary boundary, and a balance between detail and coverage was successfully achieved by the engravers

opened in 1840, and the Cranstonhill Water Works. Developing industrial patterns can be seen in the identification of pits, mills, foundries and factories, while significant public buildings and churches are marked in a heavier shade. As a balance to this, many of the proprietors of local country houses are named and similar attention is given to the mapping of gardens, pleasure grounds and parks. Once again, the representation of the southern suburbs of the city has been reduced by the inclusion of a sequence of four index maps showing the parliamentary and municipal districts, police wards and quoad civilia parishes. Martin also includes a reproduction of the McArthur 1779 plan, possibly to highlight the city's growth, which is underlined by the note of the population for 1841 as 282,134.

1844

Mapping an epidemic

Victorian Britain witnessed unprecedented social change within its cities and basic thematic maps became increasingly used to illustrate reports on such issues as health and crime. The distress of the poor during a sequence of widespread epidemics alerted many people to the deplorable filth and overcrowding which had resulted from large numbers flocking to the urban areas. In Glasgow, as in other Scottish cities, there was a considerable prevalence of fevers following the relapsing-fever epidemic of 1827–29, with over 25,000 cases admitted to local hospitals between the years 1827 and 1839. The worst year of the series may have been associated with the cotton-spinners strike in April 1837, by which eight thousand workers, mostly women, were thrown out of work. Burials in the city in 1838 amounted to 10,888, or about 1 in 24 of the whole population. This increase in mortality confounded many in the medical profession as it occurred during a period of unexampled prosperity rather than famine or distress.

Nevertheless, in 1842, acutely aware of the disproportionate numbers of deaths from various diseases among the poorest citizens, Edwin Chadwick, the public health reformer, produced a ground-breaking study entitled *Report on the Sanitary Condition of the Labouring Population of Great Britain*, which documented the need for better medical care and improved living conditions. The reports included this telling comment: 'it appeared to us that both the structural arrangements and the condition of the population in Glasgow was the worst of any we had seen in any part of Britain'. In that same year, relapsing fever, also called famine fever, reappeared in Scotland, it being first observed on the east coast of Fife and not in the crowded conditions of the larger burghs. By that summer, cases appeared in Dundee and admissions in Glasgow increased rapidly from December until about 33,000 cases, or 11.5% of the city's population, were recorded. At least 1,300 people died and it was estimated that more than

(Hugh Wilson) *Glasgow and Suburbs,* from Robert Perry, *Facts and Observations on the Sanitary State of Glasgow during the last year* (1844)

a quarter of the inhabitants of the poorer districts of the city were affected by this outbreak, which resembled cholera in its mode of progress.

Robert Perry (1783–1848), who was Senior Physician to the Glasgow Royal Infirmary, had studied the series of epidemics to affect the city in the early decades of the century and, the year following the outbreak, published *Facts and Observations on the Sanitary State of Glasgow during the last year: ... showing the connection existing between poverty, disease and crime*. Perry challenged the idea that its spread was exacerbated by the lack of a pure water supply. In his report, he commented that 'there are fewer places better supplied with water than Glasgow'. He did, however, stress the overcrowded and insanitary conditions in which many lived, where 'the darkest picture of the effects of man's cupidity is exhibited'. This report, combined with those of the district surgeons and statistical tables, gives a detailed and harrowing picture of the epidemic and the scale of its impact on the health services. Looking beyond Glasgow, the report stressed that nearly every large town had suffered and that national measures for prevention, including a poor law, were required. As with other commentators of the period, Perry emphasised a link between poverty and immorality and crime.

In order to demonstrate the correlation between disease and poverty more clearly, Perry attached this map, divided into the different city districts and marking with a darker shade the areas of highest prevalence. Black dots plotted by the city surgeons demonstrated that the most densely inhabited and poorest areas suffered most severely. The extent and progress of the epidemic was also demonstrated by monthly tables of cases for each district. Both the report and map served a further humane purpose in that they were published by the Gartnavel press attached to the Royal Asylum, opened in the year of this epidemic. The printing and colouring of the maps were done by the inmates.

It is interesting to note that the plan which Perry used was first issued by the Glasgow firm of James Lumsden and Son, publishers, engravers and stationers – although no attribution is made on the map. The original plate was engraved in 1830 by Hugh Wilson (1801?–1869), a former apprentice of Lumsden, who was to develop a successful career as lithographic printer and publisher. He is first mentioned in the minutes of the Glasgow Philosophical Society for May 1820 as 'Mr Lumsden's young man' and, by 1839, was describing himself as 'engraver and lithographer to Her Majesty'. Wilson is thought to have been the first in the city to undertake colour work. More significantly, the Lord Provost to whom Perry addressed this report was the same James Lumsden. In his remarks Perry comments, 'when you look at the appalling picture, you will employ all your energies ... in adopting such

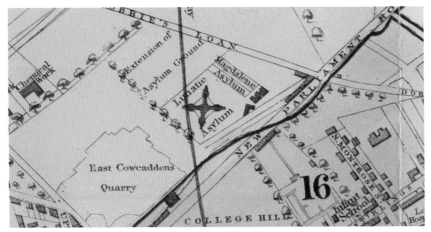

The plan continued to show the asylum off Dobbie's Loan

Sir Daniel Macnee, Portrait of Robert Perry. By permission of the Royal College of Physicians and Surgeons of Glasgow

Perry's use of black dots was an early example of statistical mapping for social purposes and, combined with shading, emphasised the concentrations of disease in the older areas of the city

measures as will be best calculated to insure attention to the wants and circumstances of the poor'.

This state of the plate identifies several new urban features, including the Necropolis, opened in 1833, the Glasgow–Edinburgh railway line, a station at Bridge Street, the proposed dock at Windmill Croft and several streets in the newly expanding west, including Elmbank Crescent and Clarendon Place. In contrast, as the plan did not cover the westward

spread of the city, it still shows the city asylum in its original location on Parliamentary Road.

Epidemics would continue to affect Britain's urban areas in the nineteenth century but gradual progress was made both nationally and locally, including the creation of a full-time post of Medical Officer of Health for the city in 1872, responding to another outbreak of contagious fever in 1869–70.

Ben Lomond

10'

Croitbanshith
Gortan
White Cairns
Thornbank

Strone, Malanoch
Hills

Doun Hill

Strone Malanoch Farm

Glenmallen House

Cruach
a Nathan

Var. 26°.50'W.

Care Loch
Head

Faslane
Bellmore

Shandon
Mein
Hill
Linhurn
Linburn
Berrydale

Letterwell
Ardinconnel
Row

Ardincaple
Castle
Helensburgh

Drumfork Cottage

Killeter Hill

Roseneath
Bay

Colgrain

Keppoch

Lower Lyleston

Ardmore Pt.

Ardardan

Geilstone

Ardmore Bay

Cardross

W. Ardoch

Clydebank

Baybank

Dumbarton

LOCH LOMOND

D U M B A R T O N

S H I R E

5'

56'

H.W. XII. 20.
Rise 9 f.t

H.W. XII. 40.
Rise 9 f.t

Dumbarton
Rock 24 f.t
Dunbuck Hill
Milltown

Dunglass Castle
Bells M.ll
Bowling Harb.r
Light F.

Donalds Quay Light E. red
W. Kilpatrick
Dalnotter

Erskine
House
Erskine Ferry

Forth & Clyde
Canal

Dalnuir Works

Newshot

R I V E R
Canal

55'

R. Cart

Blythwood
Mount

Elderslie

Brachead

Shieldhall

Linthouse

R E N F R E W

S H I R E

C L Y D E

Scotstoun

Easter Scotstoun

White Inch

Saw Mill

Govan

Partick

R. Kelvin

H.W. L.15. Rise 9 f.t
GLASGOW

50'

1846-49

The Royal Navy charts the Clyde

Despite its importance as a major estuary and a key trade route, the Clyde appears to have been somewhat neglected by earlier hydrographers. While there are several maps of the firth included in the growing guidebook and tourist literature of the early nineteenth century, these tend to be at too small a scale for navigation and certainly do not indicate the sandbanks, shoals and rocks which a mariner needs to avoid. Apart from the eighteenth-century charts by Adair and Watt, the only other published work recorded as covering these waters appears to be by Robert Blachford, a London chart-maker and publisher, who produced an enlarged survey of the Clyde in 1830.

The series of Admiralty charts which cover the firth are the work of Captain Charles Gepp Robinson, R.N. (1805–75), who was employed by the Hydrographic Office on surveying much of the west coast of Scotland, south of the Ardnamurchan peninsula. His career coincided with the period

when the post of Hydrographer to the Board of Admiralty was occupied by Sir Francis Beaufort. It was Beaufort who initiated what became known as 'The Grand Survey of the British Isles' to chart the coastal waters surrounding the United Kingdom in far greater accuracy. Probably under Beaufort's influence, Robinson was a founding member of the Geographical Society of London in 1830. Over forty of his manuscript surveys of Scottish waters are held in the Hydrographic Office Archive. From these, the printed charts were produced but as the Admiralty did not have its own facilities they were engraved by the London firm of John and Charles Walker and subsequently sold by local agents.

Entering the navy in 1819, Robinson had served on the North Sea station before joining an expedition to chart the east and west coasts of Africa under Captain Fitzwilliam Owen. He was one of the few naval officers of that mission to survive to return to Britain. Back in home waters, he worked

Charles G. Robinson, *West Coast, sheet 3. The Clyde, Loch Fyne &c* Admiralty Chart 2159 (1849)

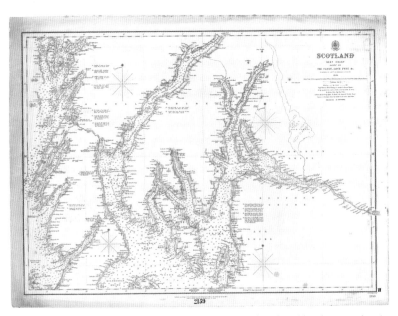

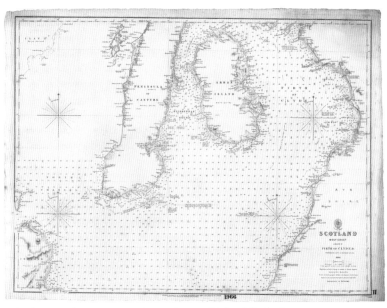

The Admiralty chart shows the same problem faced by the Watt family in attempting to include the firth and the river to Glasgow

Charles G. Robinson, *West Coast, sheet 2. Firth of Clyde &c*
Admiralty Chart 1966 (1846)

chiefly on surveying the coast of Wales before moving north to the Solway Firth in April, 1836. Initially, much of the work was conducted in hired boats and it took four years before charting the west coast proper began. During that time, Robinson was based in Dumfries or Carsethorn over the winter months, completing fair versions of his surveys. In 1838–39, a survey part of sappers under Captain Alexander Henderson, R.E. was engaged in the Clyde estuary, defining coastal points for Robinson's subsequent use. Two years later, in 1841, as part of an increased hydrographic fleet, several ships were transferred to the navy, including the Post Office steam packet *Dolphin* which was renamed *Shearwater*. She became the principal surveying ship in home waters and was under Robinson's command from 1843 until 1847 when, along with other survey vessels, *Shearwater* was sent to Ireland as part of the government's programme of famine relief.

While the complex and intricate coastline of the whole firth is clearly delineated on these charts, it is the conspicuous care with which the river itself is marked that stands as testimony to the quality of the work. They show just how

rapidly the sea bed rises off Greenock, while the shoals, banks, rocks and narrowness of the channel east of the Tail of the Bank emphasise the problems of navigation on the river itself. Much of the sounding was done using lead lines from the *Shearwater*'s boats. It is also interesting to see how the record of these soundings changed from almost exact regularity to something less regimented.

The crew of the *Shearwater* appear to have worked steadily north until, by 1849, the charting of the Clyde was complete. Robinson continued moving along the coast until, in 1851, he reached Crinan to join up with Captain Henry Otter, who had been proceeding slowly south from the northwest. Robinson was to continue in the hydrographic service until 1854, when he was posted to the Mediterranean during the Crimean War, and was succeeded by his first lieutenant, Edward Bedford. Subsequently, he was involved in the laying of the first electric cable in the Mediterranean. When he retired as a Rear-Admiral in 1870, he took up residence in Oban – a sign that he clearly had enjoyed his time in Scotland.

Robinson and the *Shearwater* were involved briefly in

another aspect of surveying. In May 1844, while still involved in charting the Clyde, *Shearwater* was dispatched to Londonderry to transfer an Ordnance Survey team and its equipment, including a zenith sector, under the command of Lieutenant William Gossett, R.E. to Stornoway to carry out trigonometrical work for the primary triangulation on Lewis. This was to provide precise location points for the more detailed subsequent topographic mapping. Only two days earlier, two midshipmen from the vessel were drowned off the north coast of the Great Cumbrae when their sailing boat was caught by a strong north-east wind. While their bodies were never recovered, Robinson and his fellow officers erected a

monument to their memory at Tomont End, which is still marked on the contemporary chart covering the Cumbraes.

Like many charts of this period, the depiction of coastal features and, in particular, towns is invaluable as these predate the work of the Ordnance Survey by almost a decade. The care with which the number of soundings is recorded on the manuscript charts gives a sense of the meticulous and arduous work involved in such an enterprise. While technology has changed since the Victorian era, the information which Robinson and his crew recorded is still the basis for today's chart.

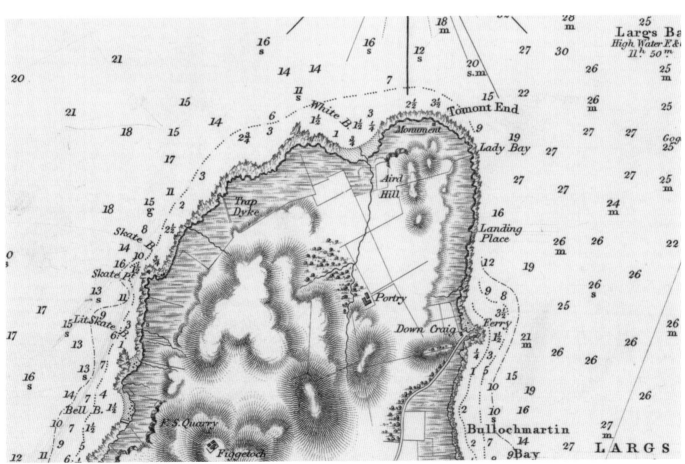

The monument erected by the officers of HMSV *Shearwater* in memory of the drowning of two midshipmen was included on sheet 1 of Robinson's chart of the Firth of Clyde in 1846

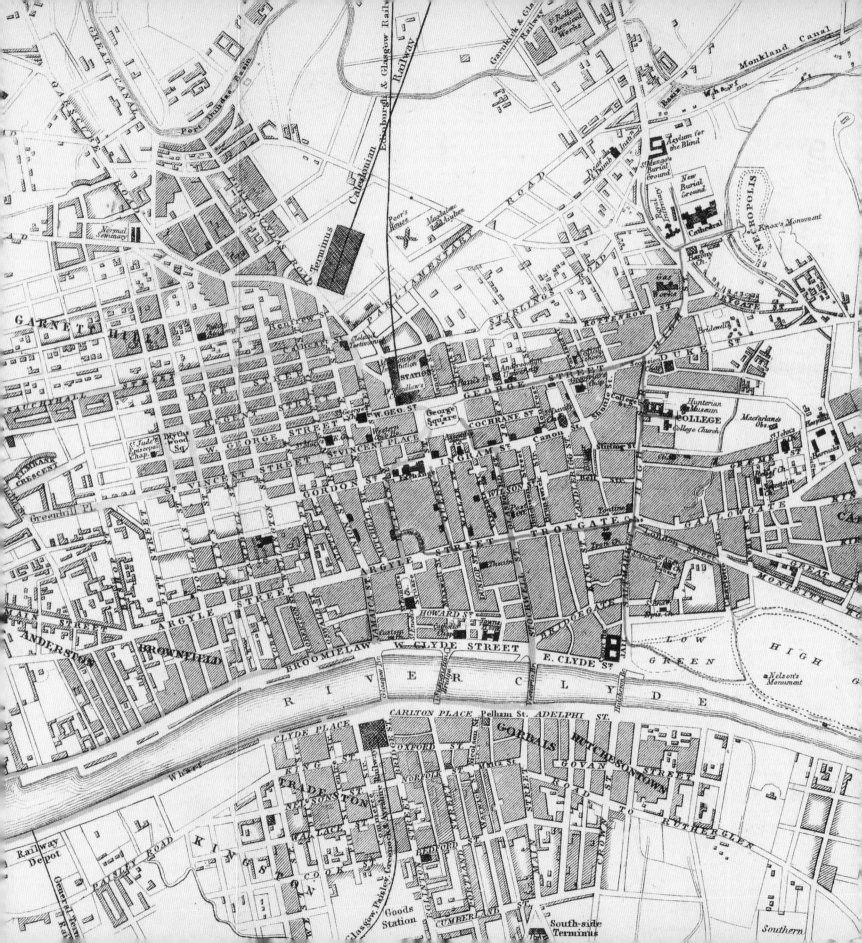

1850-60

Maps for the visitor: Glasgow as a tourist attraction

This group of three maps has been selected to show how a growing travel literature evolved from illustrated local guides into a sophisticated and extensive variety of material designed to meet an increasing tourist demand. Several of Glasgow's earlier plans were produced to accompany locally written descriptions or histories of the city. These were often designed for an audience either based in, or relatively familiar with, the immediate area. As a result of the improvements in the transport network, particularly the growing number of turnpike roads and the major impact of an extending railway system, travel steadily became easier and cheaper. Certainly, the opening of the rail link between Edinburgh and Glasgow in February 1842 resulted in an increase in guidebooks specifically including plans of the city. In addition, the increasing number of relatively inexpensive gazetteers, encyclopaedias and serial publications, which began to be issued from the mid-1830s onwards, fuelled a growing interest in knowing more about other places.

Scottish publishers responded to this trend in a range of books specifically aimed at the traveller. In 1825, the Edinburgh firm of Stirling and Kenney produced their *Scottish Tourist and Itinerary* which eventually ran to nine editions, while the first edition of William McPhun's *Scottish Land Tourist's Pocket Guide* appeared in Glasgow in 1838. Glasgow was an ideal tourist location. Apart from its Cathedral, University and other public buildings, there was good access to the Trossachs, the Clyde coast and, as the guides suggested, 'the land of Burns'. In June 1840, Adam and Charles Black advertised their new *Picturesque Tourist of Scotland* in the *Glasgow Herald*, published at 7/6d and containing engraved plans of both Edinburgh and Glasgow. The guide was described as 'comprehensive, intelligent and well-arranged' and the Blacks employed George Aikman, who had worked for William Lizars, to produce this Glasgow plan.

One of the features of this style of cartography is the small

George Aikman, *Plan of Glasgow*, from A. & C. Black, *Picturesque Tourist of Scotland* (1854)

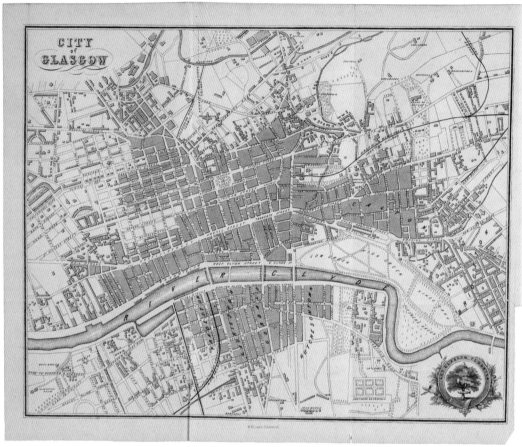

ABOVE. William Lizars, *Plan of Glasgow*, from John Willox, *The Glasgow Tourist and Itinerary* (1850)

OPPOSITE. *City of Glasgow*, from *Oliver & Boyd's Scottish Tourist* (1860)

size of the maps. These had to fit into a publication which needed to be conveniently compact and portable. The engraver's skill was in creating a map which was informative, legible and up-to-date. This resulted in another characteristic, namely their appearance in several variant versions according to the edition of the parent volume. Aikman died in 1865 and ten different states of the original Glasgow plate have been recorded up to that year but the *Tourist* itself continued to appear until its 29th edition in 1892. Another smaller plan by Aikman was engraved for two guides published in Edinburgh by T. & W. McDowall in the 1840s.

These city plans often derive from other depictions but their exact origins are rarely obvious. Engravers could be selective in what they chose to exclude or introduce and often spread their net wide in their use of sources. While the appearance, or lack, of certain features can be a guide to dating, this is not always a guarantee of certainty. This example comes from the 1854 edition of the *Picturesque Tourist* and shows how the publishers strove to provide a contemporary image. By this date, the Southern Necropolis is marked and Kelvingrove Park named, along with several of its neighbouring streets. Three coloured routes indicate walking tours encompassing the city's major buildings of note as described in the text.

Lizars himself engraved the second image selected. It accompanied *The Glasgow Tourist and Itinerary*, published by John Willox in 1850, and is based on a lithographic plan from 1847 produced by David Allan and William Ferguson. While the original displayed the city's features in a rather crude fashion, it was produced in Glasgow itself. This exemplifies the significant fact that, while these guide plans increasingly were produced outside the city, they tended to rely on earlier work by Glasgow-based engravers, as was seen in the adaption of the McPhun 1840 plan by William Nichol (**1841**). This Lizars version is, if anything, even less detailed than its apparent source but, again, introduces new elements, such as the dock at Windmill Croft. An interesting return to ornamentation is the city crest supported by both a thistle and rose – a clear allusion to Glasgow's prosperity through the Union.

The third illustration reflects the growing competition in this market and comes from another Edinburgh publication, namely *Oliver & Boyd's Scottish Tourist: guide to Glasgow*, which appeared in 1860. Oliver & Boyd succeeded Lizars as publishers of this Scottish guide and a new extended map was prepared for this edition. Although based on his earlier work, this is a fascinating combination of the old and new. Added features are particularly evident in the west of the city where a wider coverage now includes Hillhead and Partick but this has been done in a plainer style with a clear split where the extension abuts the original. Other non-Glasgow publishers, such as Thomas Nelson and John Menzies, were to produce comparable tourist guides with plans similarly based on a range of sources. Throughout this style of publishing, maps which look similar, but were not always the same, appeared in a variety of titles. Users should be cautious and avoid the assumptions of the first glance.

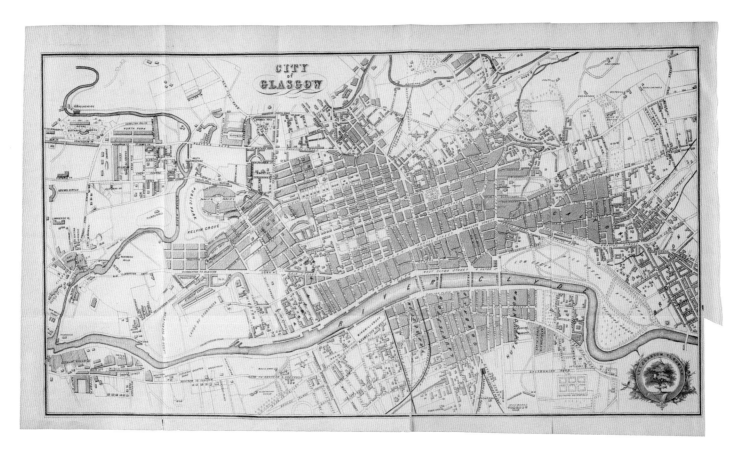

GLASGOW

ROYALTY OF GLASGOW

Possil Ho. · Cockmoor · Ballornock
High Ballornock
Eastfield · Balgray · Hillhouse
Balgray · Garroch · Keppoch · COWLAIRS STATION · Barmulloch
Laigh Bal · Kirklee Works · Newhouse · Whitevale Mill · SPRINGBURN · Laigh Ballornock
Kelvinside · Fairhill · Craigbank · Cowlairs · Petershill
Observatory · Botanic Garden · North Woodside · Hamilton Hill · Keppochhill · Flemington · Barnhill · Germiston Broomfield
North Park · SPRINGBANK · Craighall · Germiston
Dowanhill · Hillhead · Hundredacrehill · Poor House
Distillery · St Rollox Works · Blochairn
Gilmourhill · Blochairn
Wood lands · MONKLAND CANAL · Blackhill
ARTICK · Kelvingrove · STATION · Kennyhill Dist.
Ship Building Yard · Yorkhill · Cathedral · Craigpark · Kenny
P. Church · Stobcross · STATION · Golfhill · Whitehill · Haghill Distill.
Greenfield · Slip Dock · Dunchattan · Meadow park · Haghill
Broomloan · White field · College · Statefield · Annfield

Ibroxho · Plantation · STATION · Campbellfield · CAMLACHIE
Ibrox · Parkhouse · Mills · Jean field
Langshot · GLASGOW GREEN · Newbank
Bellahouston · Dockaryfauld · Newlands
Dumbreck · Gas Work · Govan Iron Works · Mills · Springbank
East Sheils · Govan Iron Works · Belvide
Hagsbooths · Sheils · Glasgow Water Companies Works
Moss Cottage · Haggs Cas. · Titwood · Oat lands · Mill · Dalmarnock
Govan Works · Coal Pits · Westthorn
Church · Yard · Dalmarnock Ho.
Campvale · STRATHBUNGO · Shawfield · Water Wks · Farme
Pollock Ho. · Camphill · Polmadie · Printfield
Knowehead · Crosshill · RUT · Eastfield Dye
CROSSMYLOOF · Hangingshaws · Hp. Ch. · STATION
Shawlands · LANGSIDE · Rutherglenmuir · RUTHERGLEN
Pathead · Battle of Langside 13th May 1568 · GI · Gallowflat Ho.
Langside Farm · M. Florida · PARK
Langside House · Clincart · Mickle Aikenhead · Laigh Crosshill
T.B. · Dovehill · Millbrae · Bankhead · Rutherglen Mill · Stonelaw
Nether Auldhouse · Papermill Farm · Mains · West Crosshill · South Crosshill
Old Steading of Newlands · Kirkwell · Aikenhead
P. Church · Holm · P. Church · CATHCART · Kennel

1852

Meikleham's map of Glasgow and its hinterland

Edward Meikleham is a rather shadowy figure in the story of Glasgow's mapping. This is somewhat surprising, given that his father, William Meikleham, was an eminent scholar. One of five brothers who attended Glasgow University, Edward is unique in being the only one with no date of his death recorded in the institution's matriculation albums. Born in 1821 to William's second wife, he enrolled in the University's Latin class under William Ramsay, Professor of Humanity, in 1834. One of his fellow students was William Thomson (later Lord Kelvin) and they became close friends. Meikleham senior had been appointed Regius Professor of Astronomy in 1799 but in 1803 he moved to take up the chair of Natural Philosophy. He subsequently held the post of Clerk of Senate and, in 1802, helped found the Philosophical Society of Glasgow, becoming its first president. Fifty years later, Edward was elected a member of the Society and he was known to mix with such figures as William Macquorn Rankine, civil engineer and

physicist, and Neil Robson, one of the city's leading engineering surveyors. In the closely knit academic community of the period, it is significant to note that William Thomson was to succeed Meikleham's father as professor on the latter's death in 1846.

There is nothing to indicate that Meikleham graduated from the University but it appears that in 1839 he entered John McNeill's engineering offices in Dublin, along with William Thomson's brother, James. James Thomson was later to become Professor of Civil Engineering at Queen's University, Belfast. McNeill had been one of Thomas Telford's chief assistants and at this time he was engaged in several railway projects in Ireland but, interestingly, he also had an office in Glasgow and was involved in the construction of the Wishaw and Coltness Railway. This clearly had an influence on Meikleham, as his only other surviving work relates to two railway commissions. In association with Neil Robson, he

Edward Meikleham, *Map of the Country for Ten Miles round Glasgow* (1852)

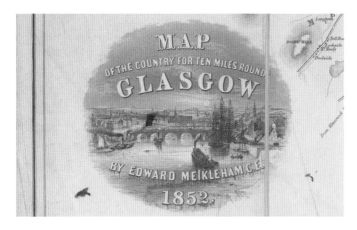

within the city and had been appointed engravers and lithographers to Queen Victoria. Prior to 1853, the Post Office Directory records their premises as at 57 Buchanan Street but, in that year, they relocated to 20 St Vincent Place. Copies of an updated edition of the map exist which show the Busby railway line extending to East Kilbride and, based on the pattern of the network, can be dated to about 1870.

Overall, this map is another example of the facility of lithography for introducing unique features to the depiction. In this case, four vignettes illustrate the corners of the map and emphasise the contrast between the industrial city and its rural hinterland. These show the busy harbour scenes at Glasgow Bridge and the Broomielaw but are balanced by bucolic images of the view of the Clyde downriver from Dalnottar Hill and at the Cora Linn waterfall with, in the foreground, a tree supporting the other three symbols of the city's coat of arms associated with St Mungo, namely the fish, the bird and the bell. Steam vessels ply the river, underlining the energy of trade, but the illustrations also emphasise that the harbour was restricted from extending upriver by the arches of Glasgow Bridge.

Like many maps of the period, it was published in both Glasgow, by David Robertson, and Edinburgh, by Oliver & Boyd. While the Edinburgh publishers have been mentioned for their production of tourist guides and maps, Robertson's interests were more focussed elsewhere. He had been apprenticed to William Turnbull, the bookseller who was involved in the production of David Smith's six-sheet map (1821), and carried on the business for seven years after Turnbull's death in 1823. He was based in Trongate and, from 1824 onwards, he published the 'Western Supplement' to Oliver & Boyd's *Edinburgh Almanack*. This association may be behind his involvement in the publication of the map. In 1837, he was appointed the Queen's bookseller for Glasgow. Five years earlier, he had published the first issue of his collections of contemporary lyrics written in vernacular Scots in *Whistle Binkie* and he is known to have been an avid collector of poetry, particularly that by the Renfrewshire poets Robert Tannahill and William Motherwell.

prepared a proposal for a rail line from the Glasgow, Barrhead and Neilston Railway near Pollokshaws to the village of Eaglesham in the November of 1852. Eighteen years later, he was working on the North British Railway's Stobcross line in Dunbartonshire.

This depiction of the city and its surrounding countryside brings to mind Richardson's map of the district seven miles around Glasgow (1795), particularly in its depiction of country houses and parishes, but, whereas the earlier image emphasises the burgh as the hub of a developing road network, Meikleham's version appears to focus more on the growing railway system for what was becoming a widening metropolitan area. Drawn at a slightly smaller scale than its predecessor, colour is used to delineate the various lines entering Glasgow from the east, south and west. Lines shown include the Barrhead, the Clydesdale Junction Railway and the Campsie Glen branch. Individual stations are named and the rather confusing mesh of lines in the Coatbridge area highlights the competition then existing between individual companies. Local estates and parklands are also identified in colour, again suggesting that the map seeks to counter an image solely given over to industry. The built-up areas in both Glasgow and Paisley are shown in block schematic form which may imply that they were not Meikleham's first priority.

The Glasgow lithographers Maclure and Macdonald produced the map and two editions are recorded. By this time, this firm dominated the production of lithographic plans

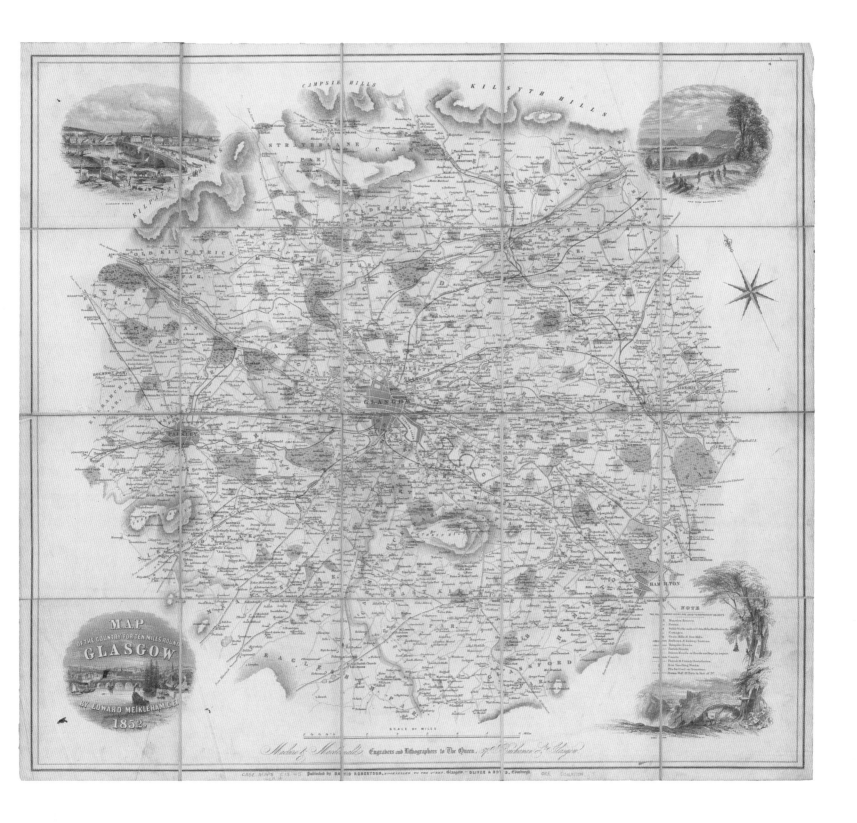

MAP
OF THE COUNTRY FOR TEN MILES ROUND
GLASGOW
BY EDWARD MEIKLEHAM, C.E.
1852

EXPLANATION.

Line of Conduit and other Works in connection with the
Loch Lubnaig Scheme projected by the Glasgow Water Works
Company, coloured GREEN.

Collecting Ground to Loch Lubnaig, shaded LIGHT GREEN.

Suggested Line of Conduit from Loch Katrine to where it would
fall into the line of Loch Lubnaig Conduit, coloured RED.

Collecting Ground to Loch Katrine, shaded LIGHT RED.

Collecting Ground to Loch Venechar below Loch Katrine, edged RED.

Suggested Conduits and Reservoirs for collecting Water in
the Valley of the Endrick, coloured PURPLE.

Collecting Ground to the same, coloured YELLOW.

R.G. *Rain Gauge.*

1853

Plans for the city's water supply: the Loch Katrine scheme

Until the end of the eighteenth century, Glaswegians depended entirely on pump wells for their water supply but, with the rapidly increasing population, such a provision soon became inadequate. In addition, the effect of sewage and other pollutants had a deleterious impact on water quality and the health of the citizens. As with the University's realisation of the impact of a growing contamination of the High Street environment, the city authorities were aware of a need for a new approach. William Harley's attempt, begun in 1804, to improve supplies by selling water from springs on his Willowbank estate was only a partial answer. Two years later, the first Glasgow Water Company was incorporated by act of parliament and, under the advice of both James Watt and Thomas Telford, it constructed reservoirs at Dalmarnock to hold water raised from the Clyde before filtration and piping into the city. Additional pump works were added but demand

continued to outstrip supply. During the 1830s and 1840s, several schemes for drawing water from a range of sources, including Loch Lomond, were investigated but none were progressed, partly due to opposition between the water companies and partly to the Council's desire to have elected trustees responsible for provision.

South of the Clyde, the Gorbals Company managed to develop a separate system for Govan, Pollokshaws, Rutherglen and, eventually, Renfrew. Gradually, the Council became convinced of the necessity to take responsibility for the supply into its own hands. In March 1852, two letters addressed to the authorities by William Macquorn Rankine, subsequently Regius Professor of Civil Engineering and Mechanics at the University of Glasgow, and John Thomson, a local civil engineer, urged an investigation of Loch Katrine as a source. They reiterated proposals, first put forward six years earlier,

John F. Bateman, *Plan shewing the District North of the City of Glasgow,*
from *Reports, on the Various Schemes for Supplying Glasgow with Water* (1853)

by Lewis Gordon, the first engineering professor appointed in the United Kingdom, and Laurence Hill. Later in 1852, John Frederick Bateman, one of the leading water engineers of his day, was appointed to examine the various proposals. Bateman had already gained experience working on Manchester's water supply and, based on his advice, as well as that of Robert Stephenson and Isambard Brunel, an act was passed in 1855, after much deliberation, in support of a scheme based on Loch Katrine as the city's water source.

As part of a wider approach to sanitary improvement, Bateman was also requested to report on the city's sewers and, in discussion with Captain John Bayly, officer in charge of the recently opened Ordnance Survey office in the city, it became clear that immediate action on mapping areas relevant to the scheme was most unlikely. The lack of an appropriately detailed city plan and a more general one of the adjacent country resulted in the Council approaching Thomas Kyle to ascertain whether or not he could supply the necessary map coverage. Circumstances changed and Kyle's services were not required.

Work on Bateman's proposals began in 1856 and, by building a small dam and raising the summer water level of the loch by 1.2 metres, more than 225 million litres of water could be supplied to the city daily. The complete system was one of the largest construction works of its day. Loch Katrine lies 55 km north of Glasgow and the system uses gravity, the water flowing, initially, through a series of aqueducts, tunnels and pipes to a reservoir at Mugdock. Since 1859, the level of Loch Katrine has been raised further, and additional reservoirs and aqueducts built.

This map is one of three which accompany Bateman's *Reports on the Various Schemes for Supplying Glasgow with Water*, submitted in 1853, and it indicates the location and possible line of conduit for the separate northern proposals put forward. While the Loch Lubnaig scheme was the preferred choice of the Glasgow Water Company, the plan also shows those based on Loch Katrine and the valley of the Endrick Water, each denoted by a unique colour. The various lochs, river systems, lines of pipe, rain gauges and hills in the vicinity are identified. Bateman's manuscript signature can be

clearly seen below the title and the plan was lithographed by George Faulkner, a typographer who had trained in Edinburgh but settled in Manchester in 1841 where he developed an extensive business.

Queen Victoria officially opened the completed waterworks in October 1859 and, at a celebratory banquet in his honour, Bateman commented that he was leaving the city 'a work which I believe will, with very slight attention, remain perfect for ages, which, for the greater part of it, is as indestructible as the hills through which it has been carried'. His vision was justified and, as most Glaswegians are proud to confirm, the city still relies on this supply, first initiated more than 150 years ago. The city's then Lord Provost, Robert Stewart, was an ardent and resolute supporter of the whole scheme and, in recognition of his services, the Stewart memorial fountain in Kelvingrove Park was erected in his name in 1871. A fortnight after the opening, the satirical magazine Punch published this 'celebratory' verse:

Glasgie's just a'richt the noo
She has gat Loch Katrine brought her;
Ever she had mountain dew,
Now she rins wi' mountain water.
Hech the blessin', ho the boon
To ilka Glasgie bodie!
Sin' there's water in the toun,
Oure eneuch to mak' its toddie.

Glasgie chiels, a truth ye'll learn
New to mony a Scot, I'm thinkin';
Water, aiblins, ye'll discern,
Was na' gi'en alane for drinkin'.
Hands and face ye'll scrub at least,
Frae ane until anither Monday,
Gif nae Sabbatarian beast
Stap your water-works on Sunday.

OPPOSITE. Bateman identified three separate schemes for the northern water supply from Loch Lubnaig, Loch Katrine and the Endrick Water respectively

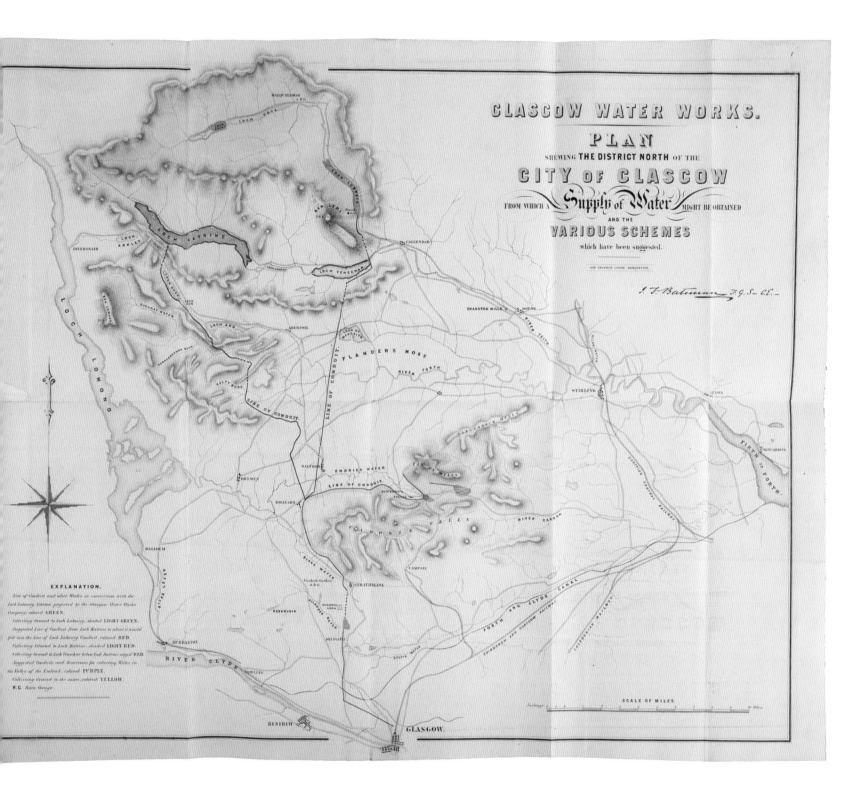

C.1856-62

The city for a wider public: atlases, gazetteers and newspapers

Not everyone could necessarily visit Glasgow but by the mid-nineteenth century anyone interested in it was well supplied with information from a bewildering range of publications. Confusingly, these often changed (or had more than one) title or had different versions of the same map appearing in individual copies of similar editions. A particular strand of such enterprise was the publication of atlases, gazetteers and newspapers illustrated with maps and frequently issued in serial format to spread costs, encourage sales and finance future work.

Robert Montgomery Martin's *Illustrated Atlas, and Modern History of the World* was first issued in this format by John Tallis, an English cartographic publisher, from about 1845. Between 1851 and 1854, a supplementary series of town plans was produced. Tallis went into partnership with his brother in 1842 and, at the end of that decade, travelled to New York, opening agencies in several American cities. By the early 1850s, he was employing over 500 staff in London, with offices in fifteen other British cities, including Glasgow, as well as those in the United States and Canada. His London Printing & Publishing Company was among the first to publish atlases from both sides of the Atlantic Ocean but his later career was blighted by bankruptcy after a failed attempt to buy the *Illustrated London News*. The individual issues of Martin's atlas appeared at a price of 1/- or 25 cents each, to coincide with the opening of the Great Exhibition of 1851 at the Crystal Palace. Varying numbers of maps seem to have been added after the initial serial publication but the Glasgow plan was first issued in part 43 (of 66) of the *Atlas*. The title page notes that the maps were prepared from government and other authentic sources.

During the 1820s, steel began to replace copper for many types of engraving. As it was a much harder medium, it could be used for thousands of impressions before signs of wear

---

John Rapkin, *Glasgow,* from Robert M. Martin, *The Illustrated Atlas, and Modern History of the World* (1855)

153

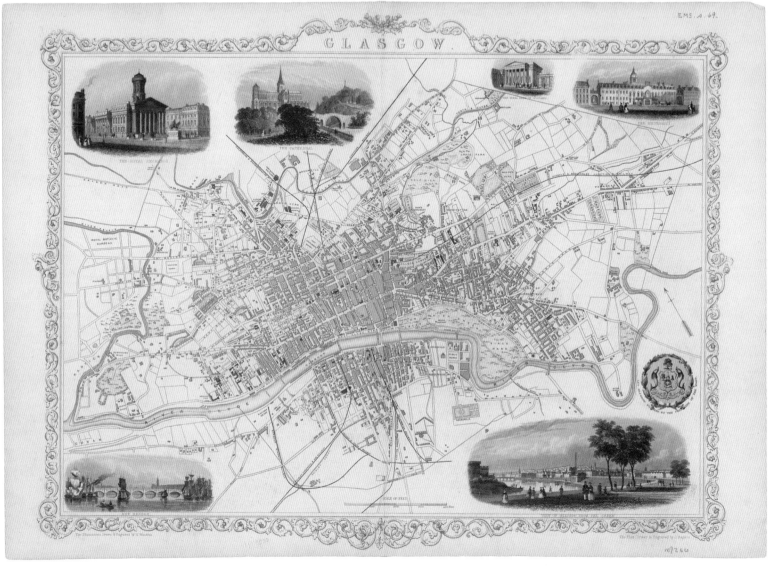

While Rapkin's work is artistically aesthetic, it disguises a poor quality in finish and errors in the naming of features

would begin to appear. Steel also allowed much finer line detail to be engraved. Along with lithographic printing, it was an ideal technical solution to the burgeoning demand for cheap illustrative material to enhance the ever-widening range of reference works then appearing. All of the atlas maps were drawn and engraved in this way by John Rapkin, principal engraver to the company, while the surrounding vignettes were the work of other artists. In the case of Glasgow, this was Henry Winkles, a leading architectural illustrator who had founded the first school for steel engraving in Germany. Due to their style, these maps are often viewed as the last flowering of British decorative cartography.

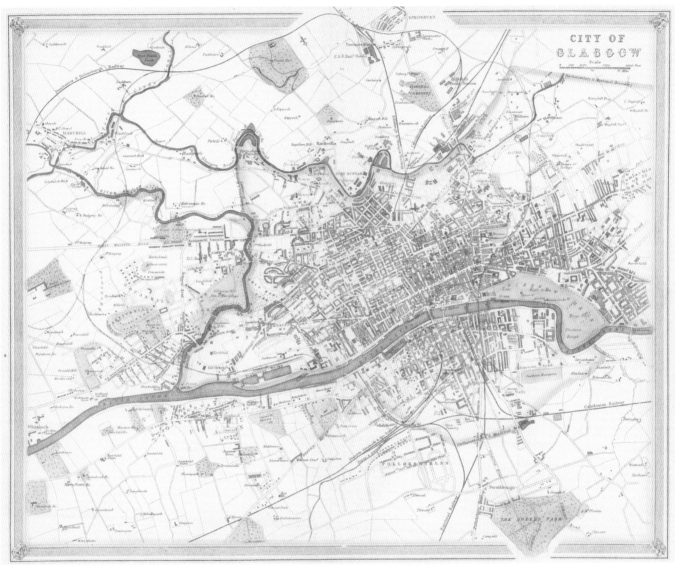

George Swanston, *Glasgow*, from John M. Wilson (ed.), *The Imperial Gazetteer of Scotland* (c.1857)

The individual plans are characterised by ornate frames and titled vignettes of prominent buildings and monuments. For Glasgow, these are the Royal Exchange, the Cathedral with the Molendinar Burn still flowing beside it, the Royal Bank of Scotland, the University, Glasgow Green and the New Bridge. Despite the esteem in which the maps are held, this plan has not been well finished. It is a reduced version of George Martin's 1842 depiction but there are several instances where the engraver's guide lines have not been erased completely, as can be seen in the 'Lands of Barrowfield'. A more serious error is the delineation of the improbably named Airdale and Monkhouse Junction Railway terminating in the

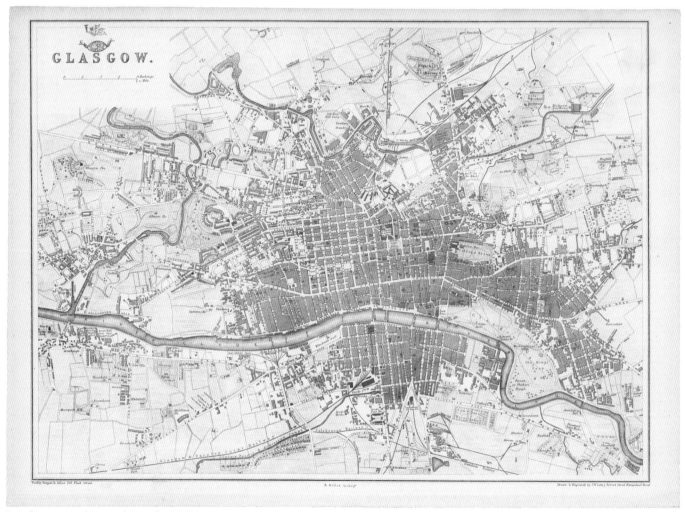

Joseph Lowry, *Glasgow*, from *The Weekly Dispatch* (1862)

middle of the College buildings in the High Street.

George Swanston's plan is representative of another strand of this mass-produced literature. This image appears in *The Imperial Gazetteer of Scotland,* edited by John Wilson and published by the firm of Archibald Fullarton & Company around 1857. The firm specialised in producing geographical works and atlases and advertised the *Gazetteer* in monthly parts or in half volumes. Swanston had a successful engraving business based in Edinburgh and drew several of the maps

accompanying this work. By 1851, he was employing four men, as well as ten colourists and his son was to become an engraver for the Ordnance Survey. Again, the city is shown spreading westward to the River Kelvin, with many new streets in this area identified. Colour has been introduced to mark railways, parks and water bodies but the plan itself has been based on a blending of the best available sources to enhance relevance and authority.

Occasionally, maps and plans were issued to accompany

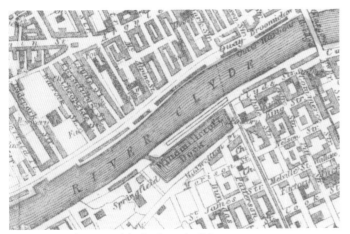

Swanston's indication of the harbour facilities

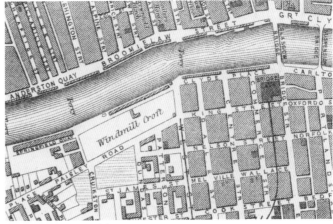

Lowry's depiction of the same

Rapkin's contrasting vignette of the genteel society of Glasgow Green

newspapers, specifically as a form of sales promotion. Between 1856 and 1862, issues of the *Weekly Dispatch* carried maps seemingly taken from lithographic plates never used for direct printing. The maps may also have been sold separately and, subsequently, they appeared together in the *Dispatch Atlas* published by George Washington Bacon. This Glasgow plan accompanied the issue of the *Dispatch* for 14th September 1862 and was drawn and engraved by Joseph Wilson Lowry (1803–79). Lowry was an English engraver specialising in scientific subjects, who produced work for the Geological Survey and the Royal Geographical Society. Edmund Weller was responsible for the lithography, the whole depiction being based on that of the Ordnance Survey but with interesting variations in the identification of such features as shipbuilding yards. As with other maps of the series, Mercury flies above a hemisphere with a scroll bearing the attribution to the *Atlas*.

1858

Glasgow's first Post Office maps

Directory maps can be quite problematical for anyone who has to deal with them. While they were produced in large numbers, frequently they were printed on thin or poor quality paper for easy folding into the front or end papers of the directory volume. Large maps required a greater number of folds which often results in a higher likelihood that, over time, the original is either torn or lost completely. In many libraries these maps were removed from their parent publication on receipt and it is not always easy to trace their history or development. On the other hand, these urban depictions can be the best record of the steady growth of a city and regularly show the gradual changes in the road and industrial patterns of the built-up area.

Joseph Swan (1796–1872) was a Glasgow engraver who appears to have begun working in the city in 1818, taking over Charles Dearie's business in Trongate. He had learnt his trade in Edinburgh as apprentice to John Buego, the engraver of Nasmyth's portrait of Robert Burns. Swan's work encompassed a range of commissions for portraits, bills, banknotes, bookplates, maps and plans, as well as illustrating the rare plants in the collection of the Royal Botanical Institution. He established his name by engraving the work of contemporary Scottish artists, such as John Fleming, John Knox and Andrew Donaldson, as illustrations for collections of views of Scottish towns and landscapes. These were published in a series of works beginning in 1828 with *Select Views of Glasgow and Its Environs*. By 1832, the *Glasgow Directory* listed him as 'engraver and publisher of the Lakes of Scotland'. In the same year, he was responsible for engraving Alexander Black's plan of Falkirk. Among later works containing his engravings can be counted William Hooker's *Perthshire Illustrated* (1843) and Archibald Fullarton's *Topographical, Statistical and Historical Gazetteer of Scotland* (1842).

His first recorded map of Glasgow, described in the

Joseph Swan, *Plan of Glasgow and Suburbs*, from *The Post-Office Annual Glasgow Directory* (1858)

Glasgow Herald as a 'neatly engraved plan', accompanied William McPhun's *Guide through Glasgow*, published in 1833. The following year, he advertised the opening of a new lithographic printing office, promoting the 'neatness, cheapness and despatch' of the new technique, which he was to use in a range of his subsequent products. A further state of the map, entitled *Glasgow New Burgh District Map*, appeared in the *Guide*'s second edition in the same year as this advertisement. Swan's focus seems to have turned to other lines of business until 1848 when a new plan was issued with the *Post-Office Annual Glasgow Directory*. This was the 21st edition of the work but the first to contain a city plan. By this date, Swan had moved his premises to St Vincent Street and the plan was to be the first in a series to be included with subsequent issues of the *Directory*. This arrangement continued up to 1865, when the Edinburgh-based company of John Bartholomew took over preparation of the cartography.

In comparison with contemporary plans prepared by other local engravers, such as Hugh Wilson, Swan's lithographic work seems cruder and provides a less detailed picture of urban features. On the other hand, the sheer prevalence of,

and continued association with, the *Directory* ensured that his plan would be a significant element in the story of Glasgow's mapping. On the map itself, Swan has reverted to a pattern of block shading for much of the city centre and makes no effort to give any sense of topography. At least eleven states of the original work exist up until 1863, showing that the map was revised to reflect the many changes to the cityscape. The *Directory* itself records alterations to the plan in the prefaces to individual years. In 1855, the addition of the West-end (Kelvingrove) Park and buildings was announced and, two years later, Swan introduced the innovation of an alphabetical index. Other earlier maps had carried lists of important buildings and main streets but Swan takes this to new levels. The lithographic process was ideal for the mass production of maps and, from this date, copies were being offered for sale at 1/6d.

This example comes from the 1858 state. Its most striking feature is the extensive list of nearly seven hundred streets and places of interest which can be located by the use of the alphanumeric grid overlay. Swan extends this well beyond a mere street directory and includes large houses, banks,

This title extract also shows part of the grid used to identify locations

Part of the comprehensive list of streets and major buildings that Swan appended to his later plans

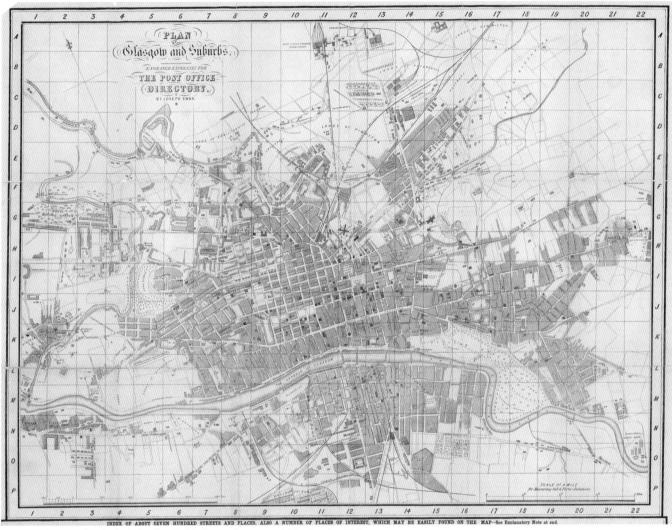

By 1857, Swan's plan included a detailed index to features within the city. This
resulted in the limitation of the depiction of the city south of the river

omnibus and coach offices, newspaper and post offices,
bridges, cab stances, hotels, libraries and public monuments;
thereby providing useful information for the visitor as well as
the resident. Its location along the bottom of the map means
that, once again, the depiction of the city's extension southwards, particularly in Pollokshields, is limited. Not all
Glasgow's streets are listed but among the new entries are
Mathieson, Great Wellington and Sandyford Streets. For the

first time on his maps, Swan has detailed the layout of the
Royal Botanic Gardens on Great Western Road, as well as
extending the coverage of Govan and Partick. One significant
feature is the scale bar which shows its usefulness for
'measuring cab & porter distances'. Along with the number
of railway lines now entering Glasgow, it suggests how far the
city was beginning to develop its own transport network.

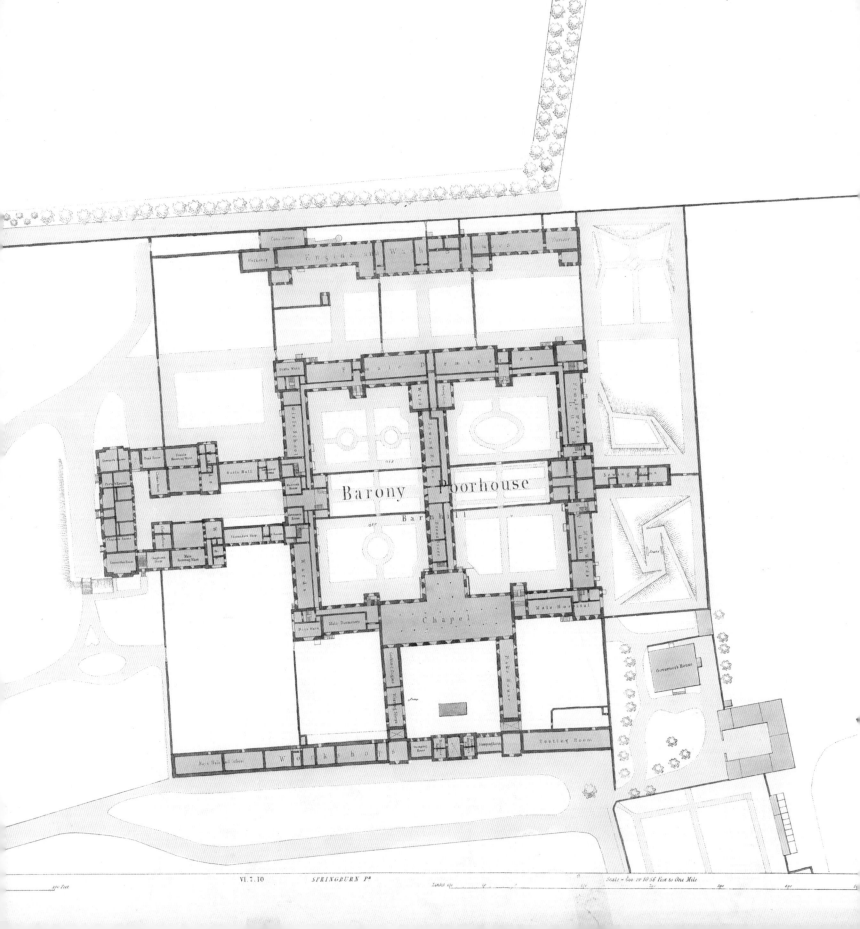

Barony Poorhouse

1859–61

The advent of the Ordnance Survey

As the national mapping agency, the Ordnance Survey has been in existence for over two hundred years. Much of its early activity in Scotland was carried out in a rather intermittent way and was completely suspended in 1823 when attention turned to work in Ireland. It was another twenty years before topographic work resumed in Wigtownshire, by then at the six-inch scale. In the interim there had been much debate and a lengthy wrangle between, on the one side, town councils, public bodies and learned societies and the Treasury on the other, on the most appropriate scales for particular purposes. Considerable lobbying occurred for larger-scale surveys to cover urban areas, since these would relieve municipal authorities of the cost of commissioning their own mapping in support of sanitary and other improvements. This reflected a change in attitudes concerning what the responsibilities of the state were, as well as a growing acceptance of the superiority of the Ordnance Survey over the work of private surveyors. A balance between national need and prudence in expenditure was nothing new and it took until 1863 before a final decision was made to accept three detailed scales, including the town plan at the scale of 1:500 for all urban areas where the population was in excess of 4,000 inhabitants.

In all this discussion, the Glasgow City Council and various other local bodies played an influential role. Contemporary newspaper reports of the Council proceedings reflect an informed and coherent consideration of the city's immediate needs. This had a particular relevance for its schemes to improve the water supply. At one stage in late 1853, the City Council considered employing local surveyor Thomas Kyle to prepare a detailed plan at a scale of one inch to 200 feet (1:2400) to show all streets and identify the levels of all crossings. Kyle felt that it was too late in the year to begin extensive surveying but thought he could have a plan finished

Ordnance Survey, *Town Plan of Glasgow. Sheet VI.7.5* (1859 61). Barony Poorhouse

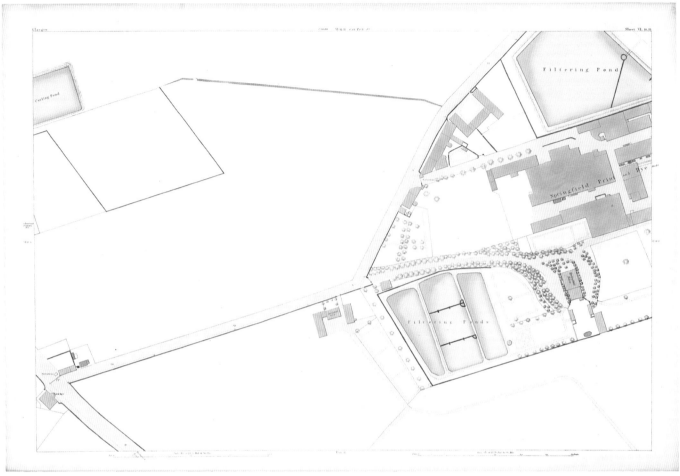

Ordnance Survey, *Town Plan of Glasgow. Sheet VI.16.16* (1859 61). Springfield Print and Dye Works

by June 1854 at a cost of £950. Events were to supersede this approach, following further discussions between the Council and the Treasury.

Although the Ordnance Survey opened an office in St Vincent Street in 1853, actual surveying of the city itself probably did not begin until 1855 at the earliest, following Henry James's appointment as Superintendent. According to its annual reports, the detailed coverage of Glasgow was finished the following year. The work itself was conducted under the direction of Captain John Bayly, R.E., an experienced officer who was later put in charge of the survey of

London. Surveying of the built-up area was conducted contemporaneously with the other large-scale mapping of the immediately neighbouring area. Following the method of the day, fieldwork in the crowded city streets was carried out in the early morning, running along either house fronts or the kerb stones of one side of the road. Once certified as ready for publication, the engraving of sheets began in 1857 and publication took place between 1859 and 1861.

The 1:500 town plan series is not the most well known of Ordnance Survey maps nor was it widely accessible before the digitisation programme of the National Library of

An extract from Sheet VI.11.12 showing details of the Fire Brigade buildings in College Street in the heart of the old town

Scotland. A single sheet covers an area of only 15.5 hectares and, as the first edition of the original Glasgow plan was published on 155 sheets, users often needed to rely on the accompanying index to locate the appropriate coverage for their study. On the other hand, as a historical record of the layout of the Victorian city, the town plan is unequalled in its detail and extent. As it was published between 1859 and 1861, it has a particular value in its portrayal of Glasgow before the City Improvement Trust began its redevelopment of some of the worst slums in the mid-1860s. However, it was not something which many could afford – a complete set of the maps of Glasgow cost £21-12-0d.

The town plan can also be used in conjunction with the 1861 census returns, the first carried out by the Registrar General for Scotland. By this date the census recorded occupations, and particular concentrations relating to district industrial specialisations can be discerned. Unlike most other Scottish burghs, Glasgow was re-surveyed at the same scale between 1892 and 1894 at national expense. It is a reflection of the growth of the city that this later edition extended to 369 sheets and comparison with the earlier version underlines how signifi-

cantly the city had changed in the intervening thirty years.

The individual sheets provide an invaluable record of the names and uses of commercial and industrial premises. They indicate the ground floor layout and internal divisions of larger public buildings, as can be seen in the sheet covering the Barony Poorhouse, while churches carry an indication of the number of 'sittings' within. As the general ethos was to support sanitation improvement, such features as manholes, water taps, hydrants and spot heights are frequently shown. Several of the sheets were coloured, following a standard pattern of blue for water, carmine for stone and brick buildings, grey for outhouses and sand yellow for roads. Colouring on the later edition was restricted to blue only. After the length of debate and the time taken to complete the publication of the survey, the final appearance of the town plan seems to have been a rather quiet affair. Commercial map-sellers had advised that newspaper advertisements were not cost effective and press notices simply list agents or areas where maps had been published. Regardless of this, the demand for the plan was so great that the Survey had to stop issuing copies from the plates until they were completely engraved.

1860

The swan-song of the privately produced plan

Based largely on information on the sheet itself, various catalogues date this map by Hugh Maclure as anywhere between 1858 and 1860 but, while it was first noted in the *Glasgow Herald* in December 1859, the announcement of publication was delayed until the following March, when it was available, coloured and on rollers, for one guinea. In April 1860, a review appeared in the same newspaper, stating that the map was reduced from Maclure's own surveys, which had taken him four years to complete, that it was engraved on steel and that it was 'very creditable as a work of art'. It is, in effect, the last detailed delineation of the city by a local surveyor prior to the work of the Ordnance Survey becoming generally available. When compared with sheets from the latter's twenty-five inch scale coverage in particular, it appears highly likely that Maclure based much of his mapping on its work. At that scale, the Ordnance Survey required twelve sheets to show the city but, while the initial survey was carried out between 1856

and 1859, publication of those sheets took another year and it was a further five years before the relevant six-inch sheet for Glasgow appeared.

Clearly, Maclure saw a gap in the market for a detailed map of the city at a scale suitable to fit on a single sheet. In this, he was part of a new trend in national cartography, whereby the information on Ordnance Survey maps was used by private publishers as the basis for commercially produced material. This can be seen most obviously in the subsequent work of John Bartholomew, largely in the series of maps produced to accompany the various editions of several Post Office directories. The first Bartholomew plan of Glasgow did not appear until 1865 and Maclure's map is, arguably, the finest locally produced city survey in the decade before the appearance of both it and the Ordnance Survey maps.

Maclure's plan is no mere copy. It was prepared at a scale large enough to indicate various public buildings, including

Hugh H. Maclure, *Map of the City of Glasgow* (1860). By permission of the Royal Faculty of Procurators in Glasgow

factories, churches, banks and foundries. In addition, it provides a more definite impression of such recent building development as Park Circus, Royal Circus and Holyrood Crescent. Like many other maps of the period, the choice of what is indicated can be somewhat eclectic. While the layout of the villas of Pollokshields is mapped, there is relatively little shown of the Partickhill scheme. Elsewhere, some of the features are still tentative, as can be seen in Woodlands and Dowanhill.

Particular attention is given to the Clyde, where unique local detail has been added in the noting of shipyard names. Cranes, ferries, quays and wharves are marked, as well as the outlines of proposed new docks on both banks of the river, at Stobcross and Windmill Croft. Upstream, the spans of the major bridges are shown, as is the weir across the Clyde constructed in association with the supply to the Glasgow Water Company's reservoirs at Dalmarnock. These and the Cranston Hill waterworks at Sandyford are both also mapped. Parliamentary and municipal ward boundaries are laid down, the latter being numbered and distinguished by different colouring

There is growing evidence of the intrusion of both railways

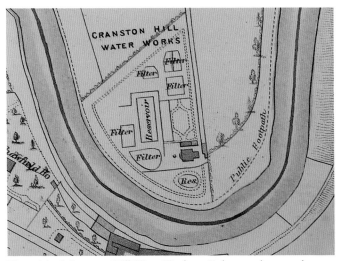

The original Cranston Hill Water Works at Dalmarnock was only one of many features associated with the water supply which Maclure identified

and industry into districts south of the river, with the identification of Eglinton foundry, the Dixon furnaces at the ironworks beside the Southern Necropolis and the Bridge Street and South Side railway stations. One interesting additional feature is the Clydesdale Cricket Club ground at the western extension of Scotland Street. Founded in 1848, it is probably the city's oldest surviving team sports organisation. In its early years, the club also played football and were the losing finalists to Queen's Park in the first Scottish Cup Final in March 1874. By then, the club had moved from this ground at Kinning Park to their present home at Titwood, the old premises being sold to a fledgling football club called Rangers. Elsewhere, the map shows the westward extension of the city beyond the far bank of the Kelvin while the gradual infill of lands in the east end out to Camlachie and Parkhead can be seen. What is also noticeable is that relatively little building is shown on land off London Road beyond Bridgeton Cross.

Hugh Hough Maclure (c.1821–92) may have started his career in the south-west. The notice of his marriage in 1847 records him as an engineer residing in Dumfries and he carried out both architectural and civil engineering work for the Glasgow, Paisley, Kilmarnock and Ayr Railway, including extensions at Kilmarnock Station and Doonfoot Bridge. Early in 1853, he set up business in Glasgow, offering his services as a land surveyor, architect and civil engineer. By this date, he claimed he had fifteen years' experience. Two years later, he was elected a member of the Philosophical Society of Glasgow. His involvement with railways is reflected on this map by a clear identification of the many lines now entering the city and the marking of the Caledonian workshops at Springburn. He was to continue this association with building designs for the Caledonian and City of Glasgow Union Companies. Subsequently, his commissions were also to include Parkhead and Rumford Street Schools for Glasgow School Board, Coats, Garturk and Bellshill Parish Churches and the extension at New Kilpatrick Parish Church. Outside his profession, he was an officer in the 1st Lanarkshire Engineer Volunteer Corps and vice-president of the Kirn Rifle Club.

This map was one of many lithographed by Maclure and

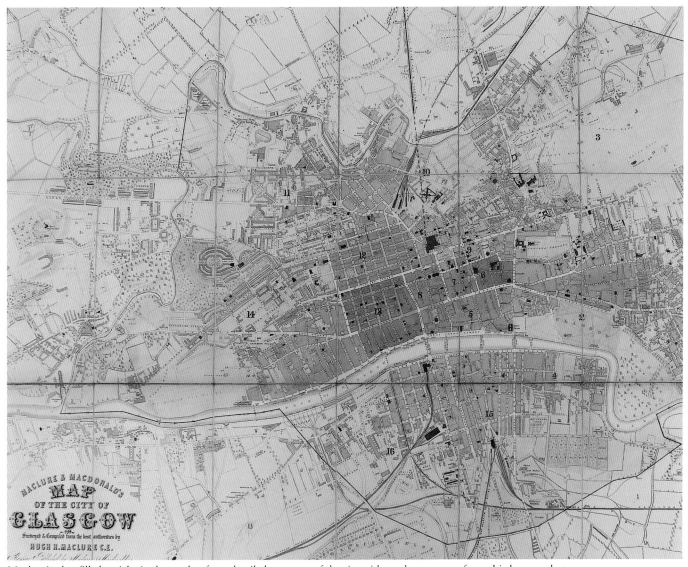

Maclure's plan filled a niche in the market for a detailed coverage of the city without the expense of a multi-sheet product

Macdonald, the leading Glasgow lithographers of their day. Andrew Maclure and Archibald Gray Macdonald formed their partnership in 1835 and the business was to extend to premises in London, Liverpool and Manchester. They were the first to use steam presses in the printing of their lithographs. Maclure himself was elected a Fellow of the Royal Geographical Society in March 1855. The possibility of a family relationship between the two Maclures is uncertain but there was obviously a close association. As the title cartouche shows, the company had, by then, been appointed 'Engravers and Lithographers to Queen Victoria'. It was to survive as a business concern until 1992.

1864

A mid-Victorian perspective:
Sulman's bird's-eye view of the city

There is a long tradition of cities being illustrated by prospect views and panoramas, beginning in Britain with Georg Hoefnagel's portrait of Oxford in 1575. As mentioned earlier, John Slezer (**1693**) had published a book of views of the most significant Scottish burghs at the end of the seventeenth century and the Edinburgh-based Robert Barker patented his invention, 'La Nature à Coup d'Oeil' for displaying his panoramic views in 1787. His depiction of Edinburgh toured with great success in both Glasgow and London, becoming a popular entertainment. Perspective or bird's-eye views, where a town was portrayed from a more elevated viewpoint, were to change dramatically in the 1820s with the advent of ballooning. While there is no evidence to confirm that any of the panorama artists did, in fact, draw their aerial views from the basket of a balloon, this type of illustration was much in vogue in the mid-nineteenth century, particularly following the first publication of the *Illustrated London News* in 1842.

It began to include large fold-out panoramic pictorial supplements, especially following the Great Exhibition of 1851. Photography was used occasionally in the compilation but most of the draughtsmanship was done at ground level.

This view of the city appeared as a folded supplement to the *Illustrated London News* for 26th March 1864 but was probably drawn in either 1862 or 1863. It is the work of Thomas Sulman, an English architectural draughtsman who worked as an engraver for Dante Gabriel Rossetti and was to prepare similar views of London, Oxford and New York. His association with the *News* was to continue from 1859 until 1888. Subsequently, he was to work as illustrator for *The Boy's Own Annual*. These hand-coloured engravings were produced with the assistance of Robert Loudan, a wood engraver said to be working in London.

Sulman's panorama of Glasgow is taken from the south and the accompanying descriptive essay compounds the belief

Thomas Sulman, *Glasgow*, from *The Illustrated London News* (1864)

Sulman's bird's-eye view encompasses an area from Gilmorehill in the west to the city's
Necropolis but, like many others, is limited in what it shows south of the Clyde

of many observers that 'all the part which is properly called Glasgow lies on the north bank of the Clyde'. Certainly, residents of the southern and eastern parts of the metropolis would be disappointed in what is displayed. An earlier aerial view of the city drawn by George McCulloch exists but is even more restricted. While the perspective of the design is not completely accurate, there is an extraordinary level of architectural detail in what is depicted. Regardless of this, the text leaves no doubt as to the city's achievement in promoting commerce. In describing the engineering improvements to the Clyde navigation, it closes thus: 'we do not know where in the world to find a better example of this sort of conquest over the difficulties of a local situation'.

Interestingly, the image presented is not one of a city oppressed by smoke or pollution but a bustling and prosperous centre. In 1861 Glasgow's population had reached over 446,000, but by the time of this view the American Civil War was beginning to have a serious impact on the cotton industry

and parts of its export trade. A key diagram and index of names accompanied the view and this identifies such features as the massive chimney of the St Rollox chemical works (then the largest in the world), the gymnasium on Glasgow Green gifted to the city in 1860 by D.G. Fleming, a native of Glasgow, several of the ferries crossing the Clyde, the Bridge-gate Free Church (which was never built) and the Theatre Royal in Dunlop Street. This had been almost completely destroyed by fire and rebuilt in 1863. While the South Portland Street and St Andrew's suspension footbridges, dating from the early 1850s, are clearly identified, the key still refers to the newly constructed Victoria Bridge of 1851–54 as the 'Old Bridge'. Apart from the many factory smokestacks throughout the town, it is the crowded scene of shipping berthed on both banks of the Clyde downstream from Telford's Glasgow Bridge (opened in 1836), which catches the eye. By 1862, more than 15,000 vessels were arriving in the harbour annually and, although the city had not yet developed any significant dock facilities, investment in extensive quays had increased the wharf area to more than 4 kilometres. Just below the Steamboat Wharf at Lancefield, Sulman indicates Napier's Dock, which was used as the only basin until the opening of the Kingston dock in 1867. It was here that David Napier, one of the most inventive of the early Clyde marine engineers, began work on the construction of iron steam vessels.

In contrast to these various signs of 'Mammon' and in accord with Victorian Christian attitudes, the view is particular in identifying the many churches within the city. The key lists no less than 37 places of worship of different denominations, excluding the Cathedral itself, but encompassing the Seaman's chapel and St John's Scotch Episcopal Church. Concern for society and its improvement can also be discerned in the identification of David Stow's Normal Seminary, the Sailors' Home on the Broomielaw and the West-End Park at Kelvingrove, opened in 1857. Given the date of its appearance in the *News*, the panorama can be compared with the Ordnance Survey six-inch map published the following year to give a greater visual impression of the contemporary life of mid-Victorian Glasgow.

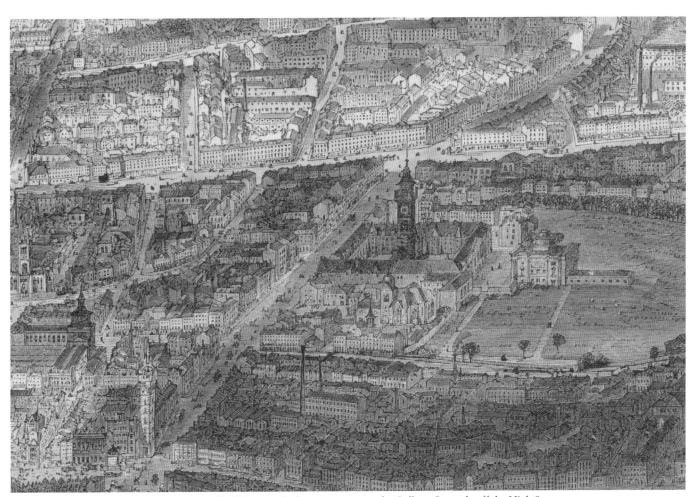

A very clear example of the encroachment of buildings and other properties on the College Grounds off the High Street

1867

Thrift and savings: the Penny banks

Single words can often have a particular connotation which defines them in terms of a certain period or idea. Thrift is one such word, for it bears all the hallmarks of Victorian values of self-improvement and a philanthropic approach to the improvement of the condition of the poor. By attempting to help raise the less fortunate out of the cycle of deprivation, poverty and crime through diligence and an element of self-sacrifice, society in general, it was hoped, would benefit. When Samuel Smiles wrote his treatise on thrift in 1875, he described penny banks as 'emphatically the poor man's purse'. By this date, they had been in existence for sixty years.

The first such bank had been founded in 1815 by James Scott as the Greenock Provident Savings Bank, but it took several decades before the working-class depositor could be encouraged into regular savings. In essence, these banks were a means whereby the poor worker could save even very limited sums to act as a hedge against hard times. Described in the words of the day, the aim was 'to promote frugal and provident habits among the humblest sons of toil'. The importance of some small savings to help cope with the vicissitudes of life was a lesson many families had learnt from bitter experience. Their money was put away not as an investment but as an attempt at a safeguard against future difficulties.

Greenock was to provide a model for Glasgow and the first similar establishment was formed in Old Wynd, in one of the poorest parts of the city, in 1850. Within ten years, 36 banks had been set up, with about 8,000 depositors. They had strong links with the savings bank movement in general and bank officials working in other institutions gave them much support. A notable example of this was William Meikle, actuary at the Savings Bank of Glasgow, who in 1871 prepared a pamphlet, *The Penny Savings Banks of Glasgow*, as a guide to their formation and management. In fact, the Savings Bank of Glasgow regarded them as an auxiliary to their own efforts

Map Showing the Penny Banks in Glasgow (1867)

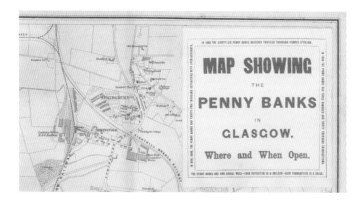

and it was a rule that, on reaching an amount of £1, the money from any account was deposited with them. By the beginning of the 1860s, the Savings Bank of Glasgow itself was the most successful in the country, with one in four of the city's skilled workmen and one in three domestic servants having an account with it.

By the start of 1866, there were a sufficient number to form a Penny Savings Bank Association and, after a general meeting, *Statistics of Penny Savings Banks in and around Glasgow for 1866* was published the following year. Figures included the number of transactions, amounts received, the number of open accounts and depositors but it also recorded the name, address and opening hours of each bank, as well as identifying the individual cashiers. The variety of banks is interesting. Within Glasgow itself, there was an Abstainer's Union and a Band of Hope penny bank. In Duntocher, the bank was opened in the Improvement Society Room, while Grove Park and Milton Iron Works both had a bank, where the intention was to encourage deposits on receipt of the week's wages. Some of the newer enterprises started in a very modest way. Pollokshaws penny bank was set up in 1866 and that year recorded only 165 transactions, with a total of only £6-7-2d. Busby, on other hand, saw more than 19,500 transactions in the same year.

It can be assumed that this map was prepared to accompany and promote the initiative of association but, compared with other contemporary distribution maps, it is a somewhat crude production. Sixty small red labels, with the names of the individual banks, their addresses and opening

hours, have been stuck rather haphazardly on a Bartholomew's base map with a pasted title covering the original inscription. Given the style, it is doubtful if the publisher had any involvement in the map's creation but, certainly, whoever did create the map used an up-to-date version of the city plan, as can be discerned by the Bartholomew address being shown as 4 North Bridge. The firm occupied premises there from 1859 to 1868. In addition, the map indicates the recently formed Blythswood Club cricket ground at Queen's Park, Great Western Road. More importantly, what the map does show is a wide spread of banks across much of the city but with a noticeable concentration in the East End and in the communities around Anderston, Finnieston and Gorbals. This contrasts with the lack of any in the city's more affluent western suburbs.

A considerable number of these banks were organised in a quite informal fashion, frequently opening for only a few hours every Saturday evening in places such as church halls, mission stations or local schools. Several of those opened in the city were specifically aimed at the young and, of the 86 banks listed in 1867, 24 were located in school buildings. Following the Education Act of 1872, which introduced compulsory education and the establishment of school boards, the Association made strong efforts to locate banks in schools and there were 14 operating in board schools by 1878.

By the time of this map, the Glasgow banks held over 32,000 open accounts and had doubled the number of receipts since 1862, despite the recent trade depression, partly resulting from the American Civil War. Fourteen years later, the number of penny banks in the Glasgow region had increased to 213 and on average they were transferring £20,000 each year to the Savings Bank of Glasgow. By the outbreak of the First World War, the figures had risen to 279 and £30,000. Like the credit unions of today, the penny bank movement belied the pre-conceived image of a uniformly feckless poorer working class.

OPPOSITE. Although the display of individual banks may appear unsophisticated, the location of labels provides an image of marked areal concentrations

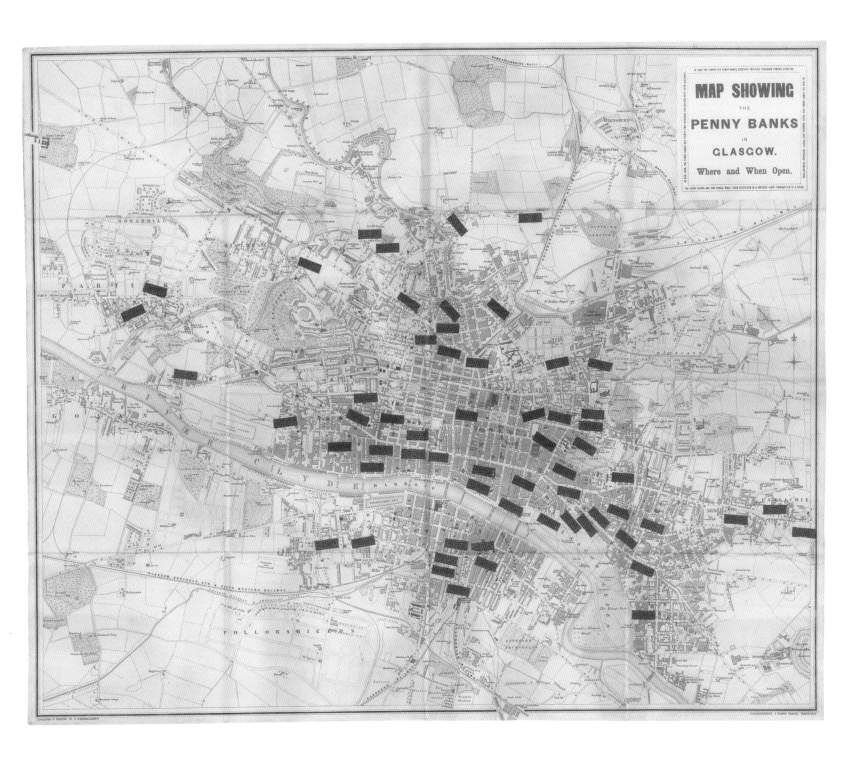

MAP SHOWING
THE
PENNY BANKS
IN
GLASGOW.
Where and When Open.

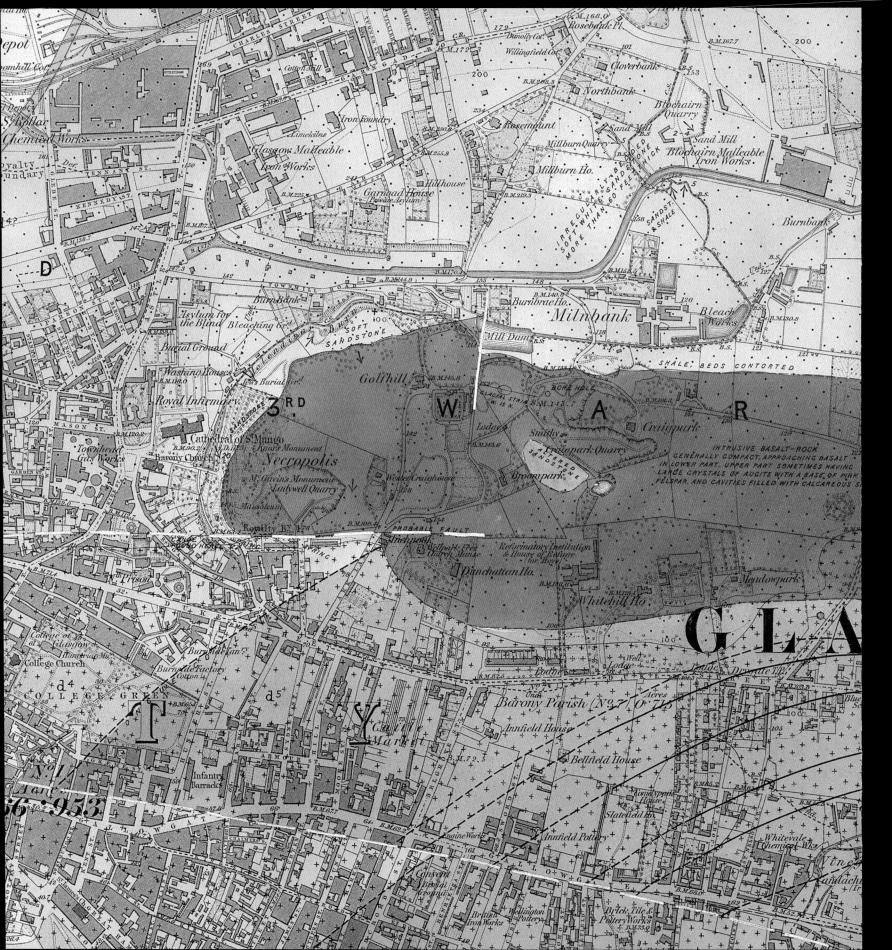

1870

Glasgow below the surface

In October 2012, the *Herald* carried a report on work commissioned by the City Council from the British Geological Survey (BGS) to map out how extensively the city was undermined by old workings. The investigation modelled the underground tunnels and shafts of an area from Paisley to Motherwell and discovered that almost half the city was standing on land with disused workings. Mining was mostly for coal and ironstone, but was also carried on to extract sandstone and limestone. Some shafts go up to 300 metres below ground. Many of the maps discussed earlier carry clear identification of several open workings which left their mark on the city. Both Buchanan and Queen Street railways stations were built on the East Cowcaddens and Provanside Quarries respectively. In fact, the arrival of the railways altered the face of Glasgow, since there was no longer a need to rely solely on local building stone and it could now be chosen from elsewhere for its quality and appearance.

The six-inch Ordnance Survey County series had been of such great practical value in Ireland that it was adopted from the outset in mapping the geology of Scotland. Publication at this scale was restricted to areas of professional or economic interest, concentrating mostly on coalfield surveys. This was greatly influenced by a Royal Commission appointed in 1866 to inquire into the United Kingdom's coal reserves. While John Geddes, a mining engineer from Edinburgh, prepared the Scottish coalfields report, other significant Commissioners were Sir Roderick Murchison, Joseph Jukes and Andrew Ramsay. Murchison and Jukes were the directors of the Geological Surveys of Great Britain and Ireland, respectively, and the Glasgow-born Ramsay was to succeed Murchison as Director in 1872.

First-edition maps were used as the base for the geological information, which was coloured by hand onto the folio sheets. This was often done by out-workers copying a master

Geological Survey of Scotland, *Six-Inch to the Mile, Lanarkshire, Sheet 6* (1870)

sheet prepared by a survey draughtsman and using selected watercolours deemed ideal for enhancing the different shades needed for the sub-divisions of the geological column. At this date, there was a constant pressure on the Survey to cover as much ground as possible every year and much time was spent on fieldwork. When confined indoors, the surveyors were involved in making clean copies of their maps and in writing memoirs on the coalfield areas. Colour printing of geological maps was not introduced until 1900–01.

This Glasgow sheet was prepared relatively soon after Archibald Geikie had been appointed Director of the Scottish section of the Geological Survey in 1867. He soon thereafter

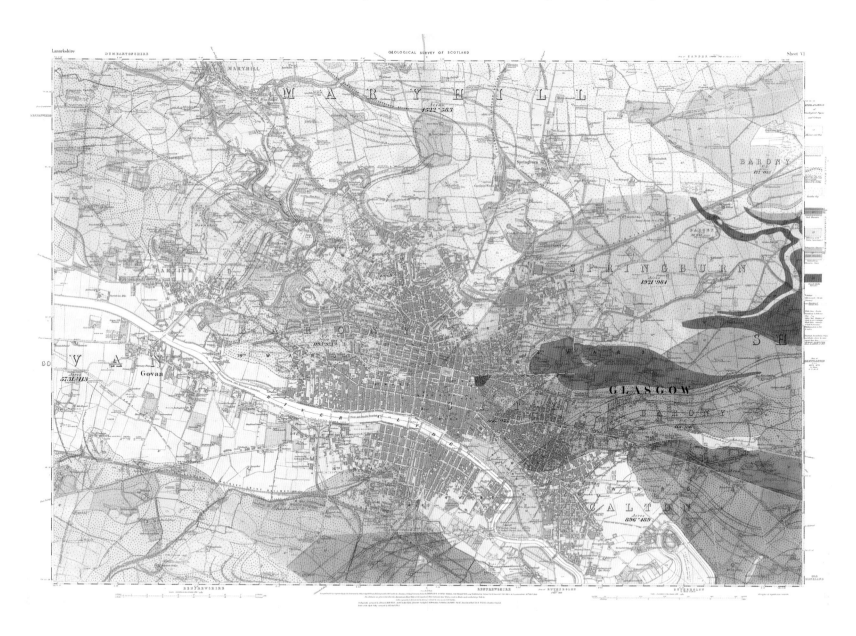

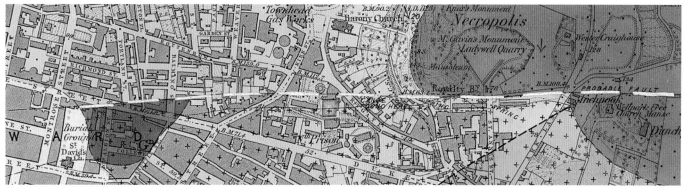

Extract showing the location of the discovery of a Neolithic canoe near Drygate

created an office in Edinburgh, allowing Scottish work to be focussed here. Edward Hull, who carried out the geological survey work shown on this map, had also served on the Royal Commission. He was appointed District Surveyor in the same year that Geikie took up his post. In effect, he was Geikie's second in command and was responsible for mapping the whole Lanarkshire coalfield. In the same year, he was elected a Fellow of the Royal Society and two years later moved to succeed Jukes as Director in Ireland. By the date of this survey, the Geological Society of Glasgow, founded in 1858, was well established. In addition, John Young, who had worked on the Geological Survey himself, was Regius Professor of Natural History at Glasgow University, a post he held until 1902. He taught university classes in geology and was responsible for moving the Hunterian collections to Gilmorehill in the year this map was published. He was to be succeeded by John Walter Gregory, explorer of the East African Rift Valley, appointed to the newly established chair of Geology in 1904.

Compared to other parts of Scotland, the geology of Glasgow was less complex to survey. Much of the city is built on carboniferous sedimentary rocks laid down about 310 to 350 million years ago, which provided it with coal and building materials. Whereas the underlying rocks affected the local form and land use, glaciers of the last British Ice Sheet flowing out from the trench of Loch Lomond sculpted the immediate area into the distinctive drumlin landscape which continues to influence the city. Much of Glasgow close to the river is indicated as built on alluvial soils, but the map also shows the extensive coal measures lying beneath the city's east end. This was covered by clays transported here by glacial action. To the west, Hull mapped out wide swathes of limestone.

The only exceptions to this uniformity are two basalt intrusions: one around North Albion Street and a more extensive occurrence east from the Necropolis through Craigpark Quarry. Across the map are marked presumed boundaries, faults, lines of dip and, intriguingly, the sites where at least five canoes were found. When the map was prepared, the trees of Glasgow's most celebrated geological feature, the Fossil Grove, were still to be discovered. One of the most famous examples of a Carboniferous forest, the fossilised stumps only came to light in 1887 during the excavations as part of the creation of Victoria Park. The trees stand in shale overlying sediments laid down by long-term flooding of the area.

While the past history of mineral extraction may cause concerns about subsidence, the Council and the BGS are investigating the use of heat energy from mine waters to help warm local homes and communities. Such a low-carbon source could contribute to making Glasgow one of Europe's most sustainable cities and provide energy for at least 100 years.

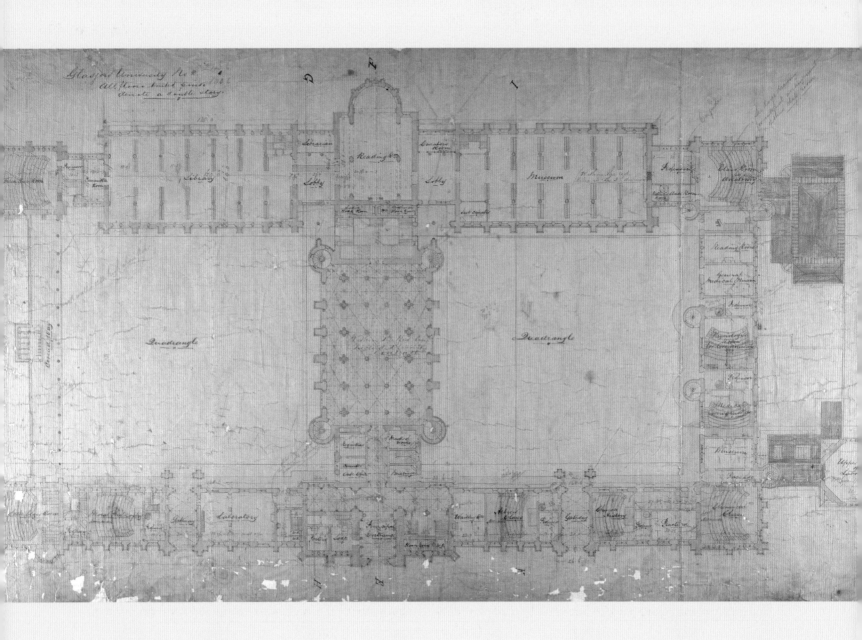

1870

The University moves to Gilmorehill

Glasgow has an impressive history in the field of education. Its High School was founded around 1124 and is recognised as the oldest in Scotland, while the University of Glasgow was established on the 7th January 1451, making it the second oldest of the four ancient Scottish universities and one of the oldest in Europe. Its creation was largely due to the efforts of William Turnbull, then Bishop of Glasgow, and, initially, it met in the chapterhouse of the Cathedral or in that of the Black Friars (Dominicans) off the High Street. For a time, it may have rented accommodation in a building later described as the Auld Pedagogy in Rottenrow before James, Lord Hamilton provided a tenement on the east side of the High Street in 1460, where it remained for more than four hundred years.

By the beginning of the nineteenth century, the environment of the High Street location was growing increasingly squalid and polluted. Incursions onto College lands by road proposals had been envisaged already. With an increase in student numbers, the professors were fully conscious of the inadequacies of their premises, both in terms of size and state of repair. Moreover, the University's finances were such that any refurbishment would have been impracticable. On the other hand, there was a body of opinion against removal to a new location on the grounds that its presence could have a beneficial impact on the residents of the older parts of the city centre.

As early as August 1845, the Glasgow, Airdrie and Monklands Railway approached the University, seeking to obtain land for a rail link into a High Street terminus. This initial scheme grew into a more comprehensive development whereby a purpose-built College would be constructed on Woodlands Hill, combined with a new hospital at Sandyford. The Glasgow architect John Baird was commissioned by the company to prepare preliminary plans which were presented

Sir George Gilbert Scott, Glasgow University. Second scheme plan. Ground floor plan marked to show the extent of the divisions of the contract (1866)

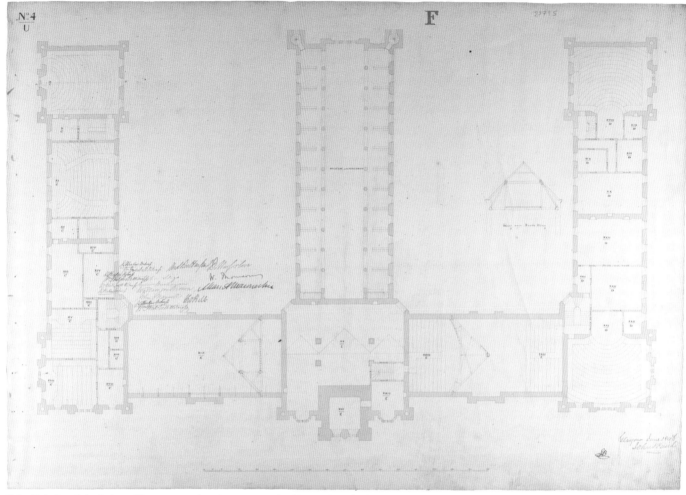

John Baird's original plan of June 1846, showing an open double quadrangle design

to the Dean of Faculty in December 1845. Baird's earlier work, such as Argyle Arcade, reflects a rather restrained design but this, his largest commission, resulted in a plan which has been described as both monumental and imposing. He produced three sets of plans, initially based on an E-shaped rectangle forming two quadrangles with an impressive main entrance, for which he was never paid. Delays in the agreement of an acceptable and affordable design eventually prevented the Woodlands scheme from being realised.

Regardless of the environmental disadvantages, the lands on the High Street were a prime location for railway companies seeking a terminus close to the heart of the city. With the drift of those with money to the western suburbs, the University's General Council increasingly supported a move but the major hurdle was financial. In 1863, the Glasgow Union Railway came up with an offer of £100,000 for the College lands. Support from the Treasury was provided and a great deal of fund-raising helped in accumulating the necessary monies. The two sides agreed on a handover at the end of five years and, in the same year, the University purchased land on the western estate of Gilmorehill.

Given such a relatively short time frame, George Gilbert Scott, the leading exponent of the Gothic Revival style, was awarded the architectural commission in 1864 without any public competition. This led to considerable local controversy. Scott was arguably the most successful man in his profession at that date and had designed several major works, including the Foreign Office, St Pancras railway station and the Albert Memorial. The construction of the new building was a massive undertaking, the largest project in Britain since the Palace of Westminster had been built to accommodate both Houses of Parliament. Work began in April 1867 and required the levelling of the top of the hill to prepare a platform for the massive foundations. There was great interest in the project's progress and the city observed a holiday on 8th October 1868, the day the foundation stone was laid by the Prince and Princess of Wales.

The final move to Gilmorehill took place in July 1870, a year later than agreed, and the official opening took place on 7th November that year. At that date the buildings planned by Scott were far from complete and it would take another eighteen years before his design, particularly the spire of the central tower and the Great Hall, was finished. On completion, the buildings provided about five times the teaching space that had been available in the Old College. Ironically, despite the greater room for classes, the new site occupied a smaller acreage than the University's previous home and the limits for expansion were to have an impact on subsequent growth.

Details of the Old College layout were to influence both subsequent schemes. Both architects designed open quadrangles, as can be seen in their diagrams. Baird's plan is fascinating, for it is countersigned by several University officers, including Duncan Macfarlan, Principal until 1857, and William Thomson. The model of two quadrangles divided by a major central feature was to re-appear on the West End site. However, the Scott version indicates an open western quadrangle, with stairs leading to the Principal's residence and the houses in Professors' Square. This was only closed with the completion of the Memorial Chapel as part of the west wing in 1929. In addition, the incorporation of the Lion and Unicorn staircase, along with parts of the gatehouse rebuilt as Pearce Lodge, were symbolic of the new being based on the old, providing a sense of both history and continuity. Change and adaptation to new methods continue to be factors in the use of all University property but few would dispute that the Gilmorehill buildings are some of the most striking in the whole city.

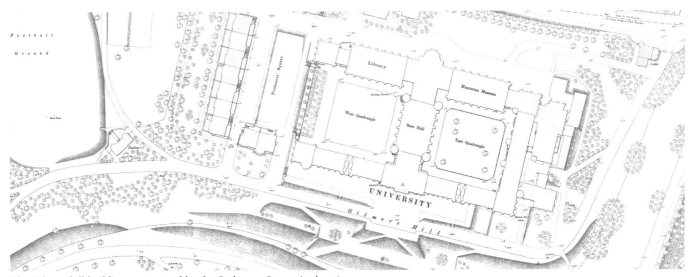

The Gilmorehill buildings as mapped by the Ordnance Survey in the 1890s

1877

The early stages of the tramway system

As Glasgow spread further and the number of people moving beyond its boundary began to exceed those coming in, the need for transport to connect new areas of population, such as Partick, Hillhead and Govan, with the shops and offices in the centre of town increased. Earlier maps (1842, 1858) indicating journey length, either by concentric rings or the measuring of cab distances, point to a requirement for transport information in some form or other. Glasgow's earliest horse-drawn omnibuses appeared in January 1845 and, as with the railways, the development of a metropolitan public transport network was led initially by private enterprise.

Following the passing of a local act in 1870, the Glasgow Tramways & Omnibus Company commenced a service between St George's Cross and Eglinton Toll on 19th August 1872. This was the start of a long relationship still close to the hearts of many Glaswegians. While the Company was granted a twenty-two-year operational lease, it was the Council which laid down the track, the ground for the first line being broken at the junction of Park Road and Great Western Road. By the final quarter of the nineteenth century, huge numbers of daily passengers were being carried. In 1893, more than 54 million passengers travelled on the Company's 300 vehicles, mostly double-deck, open-top cars drawn by pairs of horses.

Growing criticism of the service gave the Council a public relations opportunity to use the expiry of the lease to promote the benefits of local authority intervention. Ownership became an issue in the 1890 and 1891 municipal elections. Relations between the Company and the Council became strained and, in July 1894, a newly formed Transport Department took over operations. The Company intended to run in opposition and

Plan of Glasgow Corporation tramways, from Acts of Parliament and other Documents relating to the Glasgow Corporation Tramways (1877)

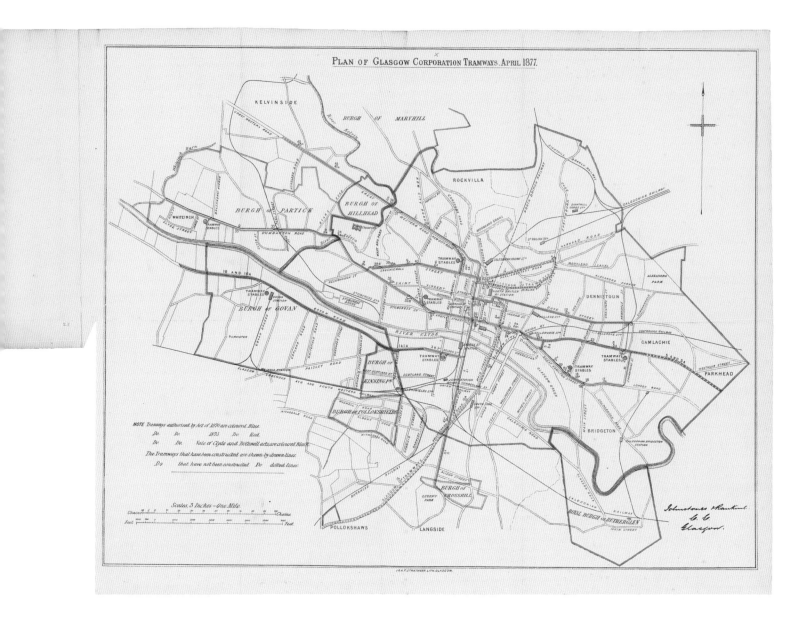

refused to hand over any of its trams, horses or accommodation. In response, the Council undercut the competition by reducing fares to a minimum. Four years later, successful experiments using electrical power were conducted and, with the imminent opening of the International Exhibition in 1901, a rapid changeover ensued. By mid-1902, all horse-drawn services had ceased. Electrification necessitated building a

separate power station at Pinkston but, more importantly, it allowed a growing number of folk to move further out. As the boundary extended, neighbouring systems, such as those in Airdrie and Coatbridge, were linked in until there was, eventually, over 150 miles of track, with lines reaching to Cambuslang, Kilbarchan and Thornliebank.

By the outbreak of the Great War, the network was

carrying over 300 million passengers and the fleet numbered almost 1,000 cars. The following year, with the impact of the numbers of men enlisting, Glasgow was the first British city to employ female tram conductresses. Some commentators have argued that the tram network contributed to the growth of the wider city but the advent of the motor omnibus serving the new housing schemes in the 1920s was equally important. These services did open up more opportunities for citizens to travel beyond the urban fringe but the reality was that the better-off gained most by cheap fares to and from the city centre.

Like the earlier illustrations of the road system, these maps indicate routes radiating from the city centre but they also emphasise the critical issue that links across the Clyde are channelled into a few key (and frequently congested) bridges. The first map shows the network in its infancy, marking lines authorised in 1870 (blue), in 1875 (red) and neighbouring systems (black). With 3,000 horses being employed, the identification of stables is an important feature. It is stated as being the work of Johnstone & Rankine, civil engineers in the city, but by this date David Rankine may have been the sole partner. He was elected an Associate of the Institution of Civil Engineers in 1874 and more than twenty years later he produced further tramway plans for the Corporation. His office in St Vincent Street is recorded into the first decade of the following century before he was joined by a family member and branched into mining engineering. James and Alexander F. Strathern, lithographers in Renfield Street, prepared the map.

In contrast, the 1916 Bartholomew map is less informative, focussing on the routes alone and indicating those in operation, those authorised (e.g. beyond Baillieston) and those where the Corporation had running powers only. It identifies neither the Pinkston power station nor the tram depots. More interesting is the depiction of the Westerton garden suburb, the first such in Scotland, begun in 1913. Tragically, the year before this was published, the recently retired City Surveyor, Alexander Beith McDonald (**1900**) died as a result of a head injury sustained in a fall from a tramcar in Sauchiehall Street.

In September 1962, the last trams paraded from Dalmar-

nock to the Coplawhill works but, in truth, the previous six years had seen a gradual reduction in the system until only the Dalmuir to Auchenshuggle route remained. Apart from Blackpool, Glasgow became the last British town to operate trams until their re-introduction in Manchester in 1992. Four years earlier, five cars re-appeared at the Glasgow Garden Festival and proved to be one of the most popular features of the whole event. While the line was only one kilometre long, over 1.5 million passengers were carried during the five months it was open. Proposals to re-introduce trams (or a light railway system) to Glasgow have been considered, without any clear final decision. Unfortunately, the financial problems which Edinburgh has faced in developing its own network are, perhaps, unlikely to encourage any step in this direction. Glasgow has a relatively low car ownership level compared to the rest of the country, with slightly less than half of all households not owning a car and, as the authority seeks to reduce traffic congestion in the city centre, the need for an efficient public transport network remains a priority.

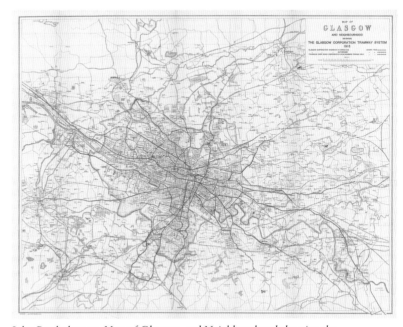

John Bartholomew, *Map of Glasgow and Neighbourhood showing the Glasgow Corporation Tramway System* (1916)

1884

The battle against the demon drink

One of the songs most closely associated with the city is Will Fyffe's 'I Belong to Glasgow', reportedly inspired by his meeting a drunk in Central Station. Glasgow's relationship with the consumption of alcohol and inebriety may be a depressingly regular cliché but, without doubt, the city has no monopoly on drunkenness. Drink and the Scottish psyche have long been a topic of debate and will continue to be as long as stereotypes remain.

John Dunlop of Gairbraid (1789–1868) is generally regarded as the 'father of temperance societies' across the British Isles, stemming from his inaugural work in Greenock and Maryhill in 1829. His work led to the foundation of the Scottish Temperance Society and by 1831 the movement numbered 44,000. A range of other groups and leagues were formed in the succeeding years, including the International Order of Rechabites, first appearing in Scotland in 1838, the Scottish Temperance League (1844), the Scottish Permissive Bill Association (1858) and the Independent Order of Good Templars (1869). A leading figure of the movement in Glasgow was Tom Honeyman (1858–1934), a prolific writer and International Secretary of the Good Templars from 1909 until 1934. His own intense passion fired his son, Thomas, to qualify in medicine but the younger Honeyman is better remembered as a distinguished director of the Glasgow Museums and Art Galleries and rector of Glasgow University.

In August 1853, parliament enacted William Forbes Mackenzie's Public Houses Bill as the Licensing (Scotland) Act, thereby regulating premises which were licensed to sell alcoholic drink to close on Sundays and at 10 pm on weekdays. This one act was to have a major impact on Scottish society until the 1970s and, in certain respects, it was to polarise attitudes towards liquor and its use throughout the

John Bartholomew, *New Plan of Glasgow . . . showing the Distribution of Public Houses, Licensed Grocer, Churches and Branches of the G.U.Y.M.C.A.* (1884)

country. While it encouraged the temperance movement to even greater restrictive action, this legislation, like the later experience of Prohibition in the United States, led to a growth in illicit drinking establishments (shebeens).

As early as March 1858, the *Glasgow Herald* carried reports and debate concerning a leviathan map, some fifteen feet by ten, created by the Glasgow Abstainers' Union to identify the numbers and location of licensed hotels, public houses and other premises in the city as part of a campaign to stem the 'tide of intemperance'. This plan was based on a similar one produced for Edinburgh. A report of a subsequent meeting mentions the identification of 1,670 properties where alcohol could be purchased, all marked on the map by a red dot and, in an echo of the earlier mapping of the 1843 fever epidemic, it was noted that the heaviest concentrations were in the poorest areas, with 90 public houses in the Gallowgate alone. Unfortunately, the map does not seem to have survived. Ayr and Elgin, however, were singled out as having the reputa-

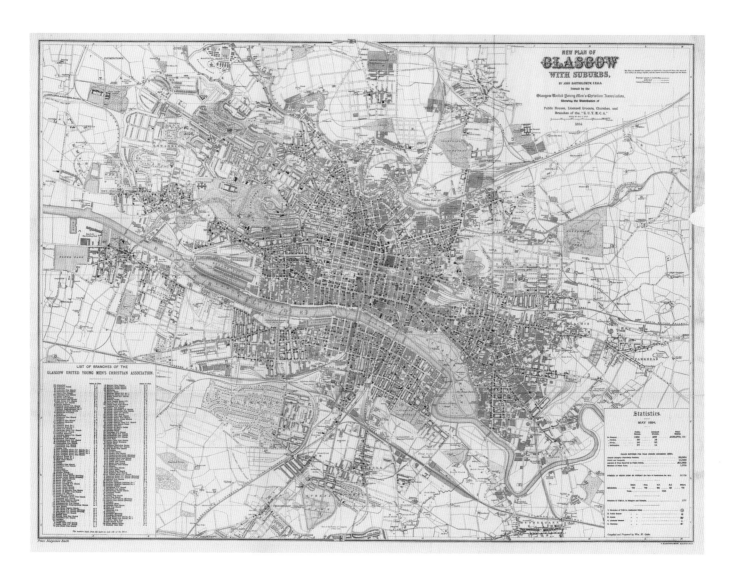

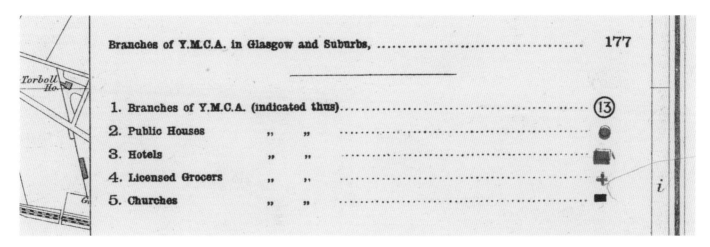

Branches of Y.M.C.A. in Glasgow and Suburbs, ... 177

1. Branches of Y.M.C.A. (indicated thus) .... ⑬
2. Public Houses „ „ .... ●
3. Hotels „ „ .... ▬
4. Licensed Grocers „ „ .... ✚
5. Churches „ „ .... ▬

tion as the worst burghs for both drunkenness and crime.

Such social reform was strongly influenced by members of various churches and the revival created by the first visit of the American evangelists Dwight Moody and Ira Sankey to the city in 1874 intensified the work of the temperance movement. They remained in Glasgow for five months and their meetings drew huge crowds to Glasgow Green. One attendee of these gatherings was a certain William Alexander Smith, founder of the Boys' Brigade in the west end of Glasgow in 1883. His ideas of discipline and self-respect were another aspect of the reforming moral zeal of the churches of the day. Temperance received a considerable boost three years after Moody and Sankey's first rallies when William Collins was elected Lord Provost in 1877. The head of a family whose company's name was memorably linked in the imagination of several generations with the publication of the Bible, he was the city's first staunchly teetotal principal magistrate, and in his time in office figures for the issuing of drinks licences fell markedly. In recognition of his public service, Collins was knighted in 1881 and a fountain in his name was erected on Glasgow Green.

This map of the city was only one of a series produced by the Bartholomew firm in support of temperance in the late nineteenth and early twentieth centuries. In essence, it is a basic plan of the city and its suburbs over-printed to identify all the public houses, licensed grocers, churches and branches

of the Y.M.C.A. However, the use of red dots to mark where alcoholic drink was on sale introduces a subtle influence to drive home the intended message of what has been regarded as a 'mass of red danger'. Issued on behalf of the Glasgow United Young Men's Christian Association in 1884, the indication of 315 churches (by small black squares) and 177 branches (by black numbers enclosed in a circle) is far less noticeable.

Certain patterns of the distribution are strikingly obvious. The concentration of premises in the older, industrialised and poorer quarters is clearly seen; a very clear example being Rutherglen Main Street. Elsewhere, particularly in the newer suburban developments of Dowanhill, Pollokshields and Hillhead, there is hardly a trace of red. The city itself is recorded as having 1,485 public houses and a further 263 licensed grocers, bringing in a total rental of more than £185,000. Taking the figures in another light, less than eighty more establishments had been licensed during the intervening sixteen years since the leviathan map was originally displayed. To further highlight the consequences of drink on society, an appended table of statistics includes annual police returns for the year, which recorded 22,364 cases of assault and disorderly conduct and 14,366 incidences of people arrested for being drunk and incapable – challenging work for a police force numbering 1,079. Clearly, there was no place for the individual who was merely 'merry' in these stark figures.

1888

Glasgow's first international exhibition

Glasgow has been home to many exhibitions and fairs over the last century and a half. The concept of large international expositions as such really began with the Great Exhibition of 1851, held at the Crystal Palace in London. An overwhelming success, with more than six million visitors and substantial profits, it engendered a string of imitations in various European and North American cities. By 1888, Glasgow, with a population of more than 750,000, was the second largest city in the realm and had a diverse industrial base which enjoyed a reputation founded particularly on its innovation. The pride it had in its achievements is well reflected in its contemporary architecture, none more imposing than that of the City Chambers, opened in the same year as the International Exhibition. As with many other aspects of corporate activity at this time, Glasgow, when it decided to host such an event, determined not do anything in half-measure.

This, the largest exhibition to take place in Scotland during

the nineteenth century, had several aims. It was a showcase for Glasgow as an industrial centre and was undertaken partly in competitive response to Manchester's success in the previous year with a similar spectacle celebrating the Golden Jubilee of Queen Victoria's accession to the throne. Like subsequent events, the city used the occasion, as the official title announced, to promote its success in the fields of science, art and industry. Manchester was an industrial rival and it was essential for Glasgow to be shown in the best light. It was also a platform to highlight the city's international interests and connections. A further objective was to raise money for an intended municipal museum and art gallery, as well as a school of art. In this, it was impressively successful in comparison with other international displays of the period. Opened on 8th May by the Prince and Princess of Wales, the official attendance was more than 5.5 million during the six months it ran. To put this in perspective, the population of Scotland at the

---

William M. Mollison, *Pictorial map of Glasgow* (1888)

195

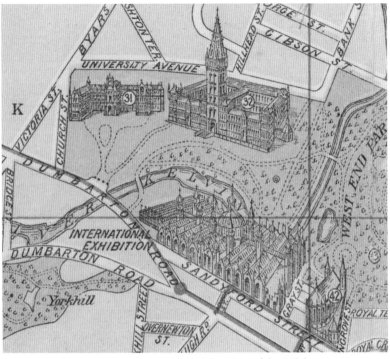

Mollison's plan gives a detailed depiction of the Exhibition
Building on Sandyford Street

British companies who used the event to advertise their wares
and several were to become household names, including
Brown and Polson, the Norwich firm of J. & J. Colman and
the successful York confectioners, Rowntree. One major
exhibitor was the London-based pottery firm of Doulton who
gifted to the city a massive terracotta fountain depicting the
nations of the British Empire. At a height of forty-six feet, it
is claimed to be the largest of its kind in the world. Recently
renovated, it now stands outside the People's Palace on
Glasgow Green.

The major displays were located in an extensive building
erected for the occasion on Sandyford Road and this is clearly
indicated on the plan. More than twenty classes of exhibits
were included in the catalogue and these covered such diverse
areas as agriculture, mining, several branches of engineering,
chemistry, foods, textiles, pottery, fisheries, education, music
and fine arts. Local shipbuilders displayed a considerable
number of models to emphasise the city's dominant position
in the industry and there were several associated working
displays of machinery. One particularly impressive new feature
was the introduction of electric light throughout the main
galleries and the 25 hectares of the Exhibition grounds, where
it was used to dramatic effect in illuminating the outside Fairy
Fountain.

While commercial aspects lay at the heart of the event,
other attractions were designed to entertain and educate
visitors. The Bishop's Castle, which had been replaced by the
Royal Infirmary, was reconstructed to display a considerable
collection of objects of historical and archaeological relevance
to the city. It was located on the other side of the River Kelvin,
close to the more boisterous switchback railway and shooting
galleries. As it was sited in Kelvingrove Park, the grounds were
dominated by the newly constructed University buildings on
Gilmorehill. Its adjoining recreation ground was used for
several of the sporting events taking place in association with
the festivities.

With the influx of tourists and visitors to Glasgow, that
year saw the publication of several guidebooks specifically
designed to show routes to Kelvingrove but also to describe

1891 census was just over 4 million. As a result of these
numbers, it realised a respectable profit of more than £41,000.

In truth, the international element of the grandiose title
was somewhat limited. Two-thirds of the exhibitors were
Scottish and only about 70 foreign companies participated.
Paris was fast becoming the major European city for such
spectacles and attracted far greater crowds. The Glasgow event
was much more a shop window for local firms, supported by
contributions from parts of the Empire, particularly India,
with which Glasgow had strong commercial links, and
Canada, which used the opportunity to promote emigration
and investment.

Many businesses in the Glasgow neighbourhood displayed
their wares, including Walter Macfarlane's Saracen foundry,
which provided an impressively ornamental iron entrance to
the machinery court, Templeton & Company (carpets) and
Coats of Paisley (cotton thread). There were many other

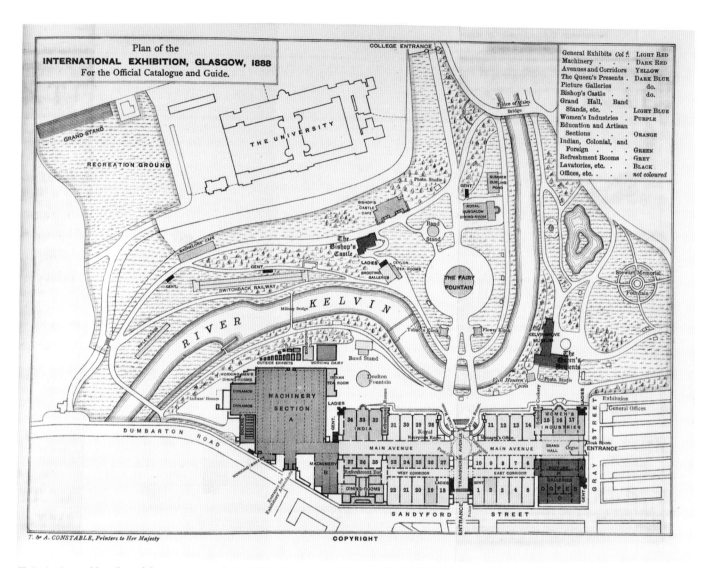

T. & A. Constable, *Plan of the International Exhibition, Glasgow, 1888*, from *The Official Guide* (1888)

tours and excursions in the surrounding countryside. The plan of the Exhibition grounds is taken from the official catalogue, published by the Edinburgh firm of T. & A. Constable, printers to both the Queen and the University of Edinburgh. Unofficial guides included those issued by the St Enoch Station hotel and local stationer and printer W. & A. Elliot. However, far more impressive was the pictorial map of the city produced by W.

M. Mollison, a Glasgow engraver, lithographer and colour printer. It clearly identifies the main building of the exhibition but fails to show any of the amusements beside the Kelvin. Like the Swan map of three decades earlier (1858), there is an extensive index to principal public buildings, streets, works and railway stations but, in recognition of the needs of visitors, it also delineates the tramway system and its stops.

1888

The city grows larger

Administrative maps rarely capture the imagination of those interested in the history of cartography, unless they are associated with a particular boundary dispute. Most of these documents are markedly functional, have minimal decoration and any use of colour tends to be restricted to the identification of ward, parish, constituency or other such divisions. They are rarely associated with dramatic events and can often seem mundane. On the other hand, like many others of their genre, they have behind them those tellingly vital elements of competence, confidence, power and control.

One of the most significant contributing factors to the increase in Glasgow's population in the later nineteenth century was the extension of the city boundary. By 1861, the city itself was experiencing lower rates of population growth, while subsequent decades saw a net migration to such neighbouring districts as Partick and Govan. This outward movement has already been mentioned as being in progress since the second half of the previous century but it was fuelled increasingly by the combination of a developing railway network and the greater availability of properties with affordable rents elsewhere. This created more than a change in population structure since the loss of the more affluent to another authority had a serious impact on the city's finances.

By the middle of the nineteenth century, there was a growing awareness in Glasgow of a fundamental fact which continues to be an issue in civic government, wherever that may be, to this day. This is the question of suburban residents enjoying the facilities and amenities which a major regional centre has to offer without contributing to that city's revenue through whatever local taxation is in place. Glasgow's neighbouring communities included nine Police Burghs who jealously guarded their independence and were wary of what

John Bartholomew/House of Commons, *Plan of the City of Glasgow and the Surrounding Districts*, from *Report of the Glasgow Boundaries Commissioners* (1888)

was seen as Glasgow's aggrandisement. Lord Provost Andrew Orr is often credited as the leading proponent of the vision of a 'Greater Glasgow' and, undoubtedly, he was convinced of the need to extend the city's boundary to incorporate those more affluent communities into a wider municipality. Orr was in office during the mid-1850s at a time when Glasgow was making major efforts to erase the stigma of bad housing and health which Edwin Chadwick had highlighted in the previous decade.

A developing civic administration required a significant source of revenue to support it and there was a coordinated drive to establish Glasgow as a reassuringly competent authority. Several of the plans previously discussed are indicative of an increasing involvement by the municipal authorities in the provision of services, such as water supply, and in what may be termed environmental improvement in general. In part, this was due to force of circumstances as much as to any determined drive towards public ownership. Nonetheless, the scale of many of such projects required a level of organisation and expertise which the Council was well placed to provide. As the city's participation in activities which were once the domain of private enterprise grew, so did the reputation of its officials for effectiveness and professionalism.

Much of the subsequent increase in the city's administrative domain was overseen by the formidable figure of James D. Marwick, Town Clerk between 1873 and 1903. Trained as a solicitor, Marwick had held the same post in Edinburgh before being offered almost three times his salary to move to the west. In many respects, he was the epitome of municipal assuredness. An expert on Scottish local government law, his opinion was sought by other burghs and successive Lord Advocates. Under his guidance, the city built up a considerable case for boundary extension and leading historians of Glasgow regard his 'tenacity and determination' as key features in the considerable expansion of the city in 1891.

Eventually, the arguments for a more cost-effective unit succeeded. Between 1872 and 1896, three separate acts annexed a number of peripheral suburbs which were, in reality, part of the growing city. Kelvingrove and the estate of Gilmorehill were engrossed in 1872, along with Kennyhill, and this established a precedent for the future. Nineteen years later, the addition of such areas as Govanhill, Hillhead, Maryhill, Possilpark, Pollokshields and Langside was to add 53,000 people to the number of residents. In 1885, a Boundary Commission report in relation to the redistribution of parliamentary seats specifically mentioned Glasgow's pressure for an increase of the municipal jurisdiction. While the Commission's emphasis was on contiguity, such extension was seen as beyond its remit and, generally, any territorial increase was viewed as likely to impact on the balance of political power. One of the commissioners was General John Bayly, who had been the officer in charge of the original Ordnance Survey of Glasgow. An accompanying plan indicated four small proposed extensions for the city in Gilmorehill, Alexandra Park, around Coplawhill and north of St Rollox.

The plan here presented accompanies the 1888 Glasgow Boundaries Commission Report which recommended the extension implemented in the City of Glasgow Act of 1891. This increased the total city area to nearly 5,000 hectares – almost doubling the authority's domain. This is clearly depicted on the plan where the proposed future city is marked in colour, along with the new boundary recommended by the Commission. It is significant that the depiction uses a Bartholomew base, showing that the company was trusted to provide a reliable map for central government purposes. The Report itself indicates tensions between ratepayers, who supported annexation, and local commissioners, who were opposed to it. Individual districts were investigated and the need for proper financial support for services underlined the case for incorporation.

Despite the opposition put up by some burghs, there seems to have been a certain degree of inevitability in the process of annexation but, once more, a careful study of the plan shows a few areas, such as Partick, Govan and Pollokshaws, which held out independently for a little longer.

OPPOSITE. The whole plan indicates the considerable extension of the area of the city to the west, south and north-east

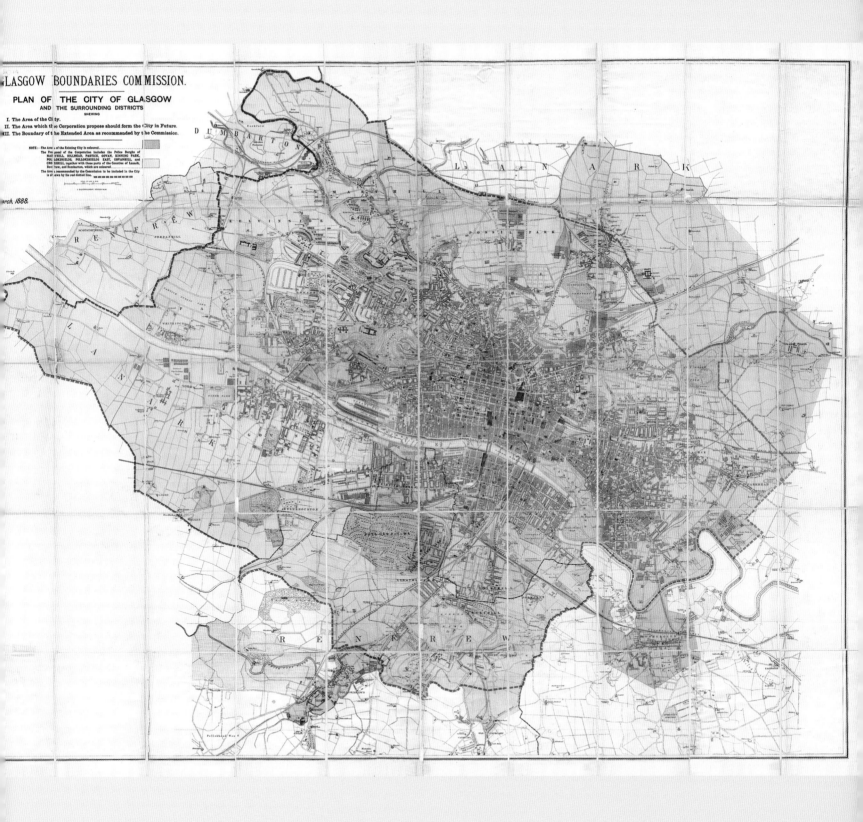

GLASGOW BOUNDARIES COMMISSION.

PLAN OF THE CITY OF GLASGOW
AND THE SURROUNDING DISTRICTS
SHEWING

I. The Area of the City.
II. The Area which the Corporation propose should form the City in Future.
III. The Boundary of the Extended Area as recommended by the Commission.

NOTE:—The Area of the Existing City is coloured........
The Proposal of the Corporation includes the Police Burghs of
MARYHILL, HILLHEAD, PARTICK, GOVAN, KINNING PARK,
POLLOKSHIELDS, POLLOKSHIELDS EAST, GOVANHILL, and
CROSSHILL, together with those parts of the Counties of Lanark,
Renfrew, and Dumbarton, which are coloured
The Area recommended by the Commission to be included in the City
is shewn by the red dotted line.

March, 1888.

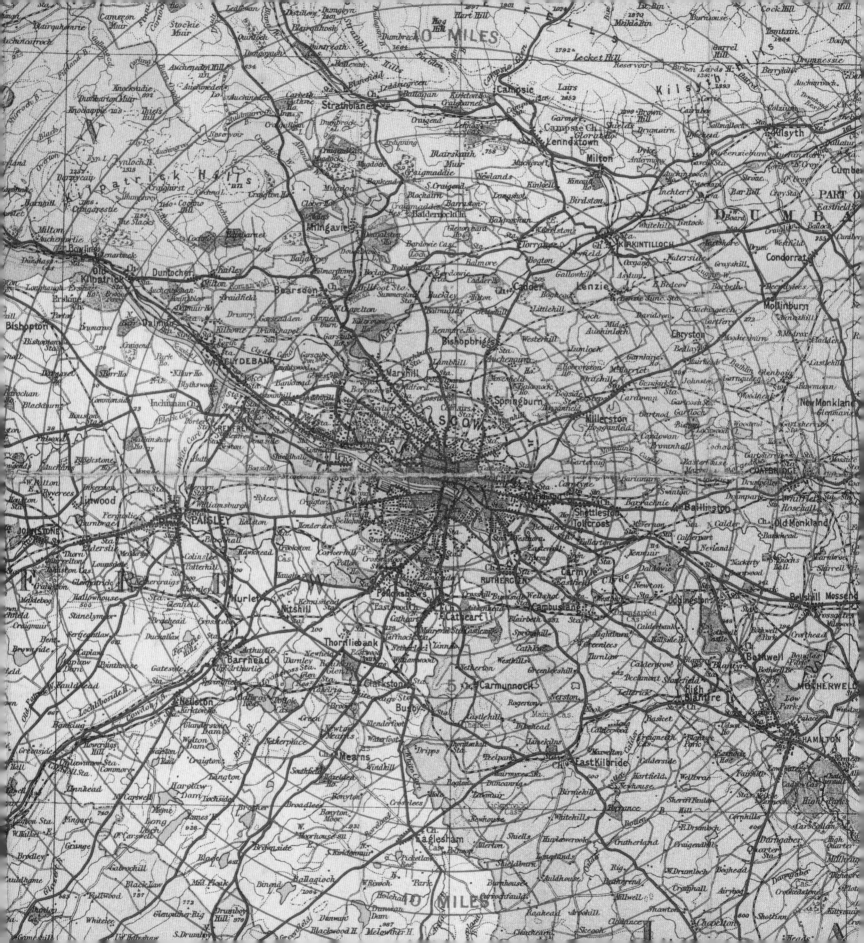

1895

When the bicycle was king

In the year that W. & A.K. Johnston engraved their first major plan of the city (**1842**), some Glasgow residents witnessed their first bicycle accident, when a Dumfriesshire man, riding what was described as a velocipede, knocked down a child after being surrounded by a large, interested crowd. It is difficult to ascertain whether this cyclist was Kirkpatrick Macmillan, popularly acknowledged as the 'inventor' of the bicycle. However, while it was only one of the many experimental stages in the evolution of this mode of transport, none achieved much popularity and they were frequently met with opposition from both pedestrians and local authorities – in this case, the cyclist was fined 5/- by the Gorbals police.

Only with the development of the 'safety bicycle' was a wider public reached and cycling became popular. The appearance of the Rover bicycle, produced in Coventry by John Kemp Starley, combined with the introduction of pneumatic tyres in the late 1880s, marks the transformation in the impact of cycling. Further changes, such as the diamond frame shape and geared chain-driven back wheels, reduced the likelihood of riders falling off but, equally, made the bicycles lighter and cheaper. At a time of growing consumer demand, advertising was used to stress the possibilities of a new sense of freedom and independence that ownership would allow. This was particularly effective in appealing to women.

This map dates from the high point of what is often termed the 'bicycle craze'. By 1896, it is estimated that Great Britain had more than one and a half million cyclists. For the first time, town dwellers could explore the immediately surrounding countryside relatively cheaply. Cycling provided an escape from the commonplace of the city and, inevitably, a cycling fraternity became established. Glasgow had fifteen cycling clubs by 1888, with more than another forty being formed before the end of the century. Weekend excursions to areas such as Drymen, Strathblane and Fenwick became part

W. & A.K. Johnston, *Cyclists' Touring Map of Glasgow and Surrounding Country* (1895)

of the season. As a result of the growing interest, there was an increase in the publication of material devoted to cycling, in particular periodicals, road books and maps.

Many commercial map publishers seized the opportunity to offer products specifically geared to this market. These included George Philip, G.W. Bacon and, in particular, the Edinburgh firms of John Bartholomew and Gall & Inglis, whose contour road books with continuous strip maps were ideal for touring. It was rare for these to be new surveys and they frequently were based on reductions from Ordnance Survey publications or from existing stock, with the substitution of new titles. The best maps were those which indicated road quality, hill gradients, repair workshops and places where

the traveller could find refreshment. Others were produced solely to aid the more adventurous in route-finding over a wider area. At this time, the Ordnance Survey did not produce coloured maps at scales which would meet the needs of these particular consumers. From 1893, the Glasgow publisher Hay, Nisbet & Company produced *The Scottish Cyclist Road Book and Annual* as a concise reference book, containing descriptions of popular routes, a hotel directory, lamp lighting tables and other useful information.

While this map is described as a *Cyclists' Touring Map of Glasgow and Surrounding Countryside*, it provides little in the way of relevant detail, apart from the sequence of red concentric rings showing the distance in miles from George

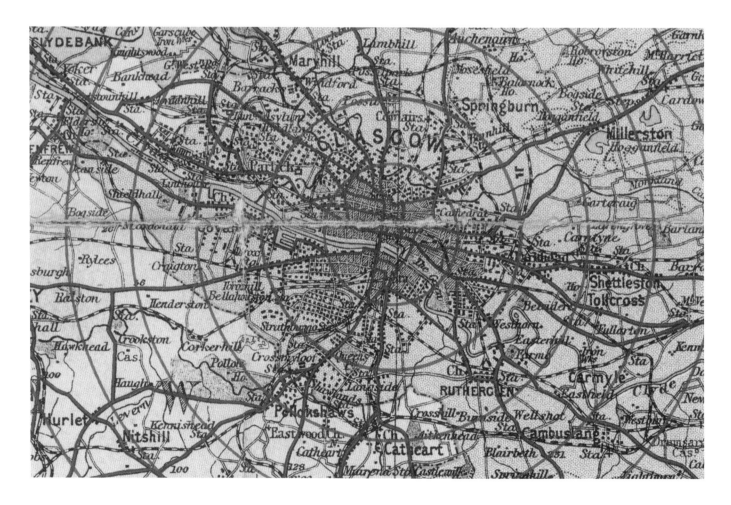

Ayr, the final destination of one of the cycle tours in an accompanying booklet

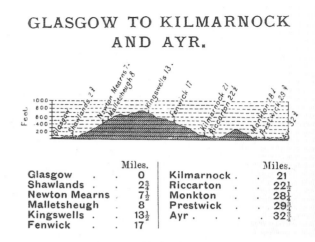

A gradient profile and distance table for the same route

Square. This change from Glasgow Cross is, in itself, significant. The map gives no indication of road quality, terrain or gradient and merely replicates the information from the Johnstons' own *New Map of Scotland for motor, cycle and pedestrian touring and general reference* published at the same scale, in sixteen sheets, of about the same date. Colour is used only to identify the individual counties on the map and there is no significant effort to differentiate roads. However, the map is accompanied by a booklet of 21 cycle tours with descriptions of both roads and local objects of interest, as well as contour illustrations of the gradients. This example (above) covers the route to Kilmarnock and Ayr and the description, with several quotations from *Tam o'Shanter*, ends in Alloway.

By the time this Glasgow map was produced, both the company's founders were dead but another brother, Thomas Brumby (1814–97), was in charge of the firm, now based in its new Edina Works, in Edinburgh. He had worked on the *Royal Atlas of Modern Geography*, had opened a London office in 1869, and was a founding member of the Photographic Society of Scotland. In March 1877, he was appointed Geographer to Queen Victoria for Scotland and this was extended in 1886 to the post of Engraver and Copperplate Printer.

Although the company was technically proficient and specialised in maps for schools, particularly wall maps, many of its atlases and series were never as popular as the competing Bartholomew products. Bartholomew's notable success was in the representation of height by layer colouring, making their half-inch series clearer and more legible, and eventually making it the commercial market leader.

One of the most limiting factors in the enjoyment of the 'open road' was the poor condition of many road surfaces but, unlike other parts of the country, their higher quality in Scotland was a contributory factor in the development of cycling north of the border. Within a decade of the publication of this map, however, the cycling boom was over as interest turned more to motor cycling and, then, automobile touring. Nonetheless, major cycle manufactories were established in the city at this time, including the Victoria Company in Dennistoun, which promoted itself as the largest Scottish manufacturer, and Howe in Bridgeton. Five years after this map appeared, David and Agnes Rattray began business in McAusland Street in the city, laying the foundations of a firm which would develop into one of Scotland's enduring cycling landmarks.

1895

The widening suburban rail network

In 1842, James and Robert Hedderwick, sons of the Queen's printer, established the *Glasgow Citizen* as a weekly newspaper, featuring the work of contemporary Scottish authors. James himself was a minor poet and contributed work to other journals. Twenty-two years later, he was to commence publishing the *Evening Citizen* as the city's first daily evening paper, priced at one halfpenny. The success of this enterprise gave it the largest circulation of any in the west of Scotland. By the date of this map's appearance, James Hedderwick had retired and his sons were running the family printing and publishing business, now based in its own building in St Vincent Place. Apart from the title, this map is further associated with the firm by the marking of distances from the *Citizen* Office, identified with a black star, by mile interval circles.

This *"Citizen" City and Suburban Railway Map* dates from 1895 and was first advertised in the *Evening Citizen* at the end of March as 'the cheapest map ever published'. In a quarter-page notice, the publishers emphasised the care with which it was prepared and corrected and Bartholomew himself is quoted as to its value. Priced at only 1/-, the map focusses on the railway network of the metropolis, indicating a dynamic system at the height of its impact on the fabric of the city. By using colour to identify the rival lines, a clear image of the individual companies (North British, Glasgow and South Western, Caledonian and City of Glasgow Union) and the network is provided. Routes are named and those lines under construction are identified, including the Cable Subway (as it was then known) and its stations. As it had not yet been completed, two of the stations have been named, rather unimaginatively, as Partick East and Partick West. One particularly noticeable feature is the recently finished Cathcart Circle, the south side's most successful suburban rail venture. The map was clearly designed for the consumer, as it illustrates

John Bartholomew, *'Citizen' City and Suburban Railway Map of Glasgow* (1895)

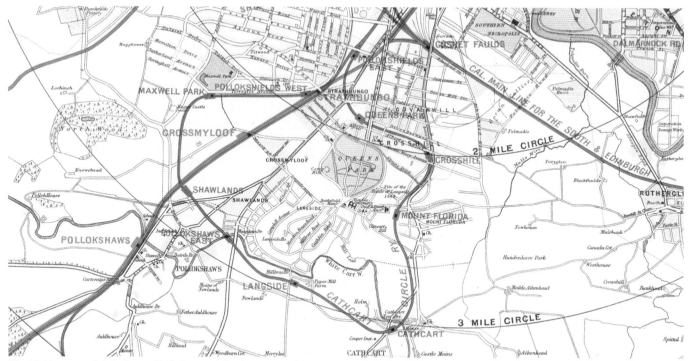

The highly successful Cathcart Circle line and the distance circles showing little urban development outside its loop

the passenger lines, identifying more than 130 stations, whereas there is no colour indication of, for example, the High Street College or General Terminus Quay Goods stations.

Elsewhere, the Glasgow Central line linking through to the Lanarkshire and Dumbartonshire highlights the Caledonian Railway's efforts to claim a share of business on the north bank of the Clyde, in competition with the North British Company. Within the city itself, this was to involve an underground connection through Central Station, a project fraught with construction difficulties, requiring some demolition, and considerable compensation. The line was opened finally in August 1896 and gave the Caledonian a considerable boost in accessing suburban traffic. Such competition lies at the heart of the often confusing plethora of tunnels, lines and stations which can still be found in Glasgow. What is possibly more informative is the lack of any significant impact on the geographical spread of the city at this date. This is particularly

true in areas to the north-east, east and south-west and can be exemplified by the marked lack of urban development between the three and four mile circles.

One of the key features of the rail network in the Glasgow area continues to be the lack of a through link from north to south. Main line passengers travelling beyond the city cannot cross the Clyde without transferring by foot or service bus from Queen Street to Central Station. For many years, any scheme for a rail bridge across the river met with opposition from rival companies, the Town Council, River Trustees and Bridge Trustees. Such a crossing did not come until 1870, with the creation of the City of Glasgow Union Railway. Despite being a relatively short link from Pollok to Springburn, it also faced both major construction difficulties and expense in land purchase. The line originally terminated at a station in Dunlop Street but, eventually, St Enoch Station was officially opened in October 1876 as a fitting welcome to passengers arriving

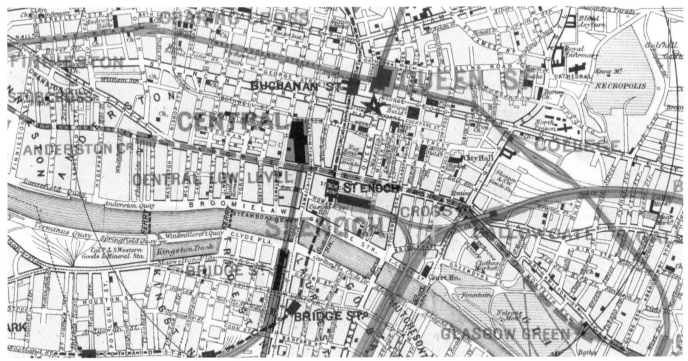

The density of railway incursion into the city centre is clearly seen in this extract, as is the star marking the offices of the *Glasgow Citizen*

from the south. Three years later, its imposing associated hotel opened its doors to travellers, the first major public building in Glasgow to be lit by electricity. The Caledonian Company had taken little interest in this line and, for thirty years, pursued a scheme of its own, which came to fruition with the opening of Central Station in 1879. Initially, it was a poor compromise but, with a re-design between 1901 and 1905, the city gained not only an impressive major station but also the 'Heilanman's Umbrella', that sheltered part of Argyle Street created by the overhead rail bridge.

Several of the competing lines and stations were never profitable and closures occurred throughout the twentieth century. With the rationalisation of rail services in the 1960s, the city centre was left served by only two main line stations, following the closure of St Enoch and Buchanan Street. While subsequent improvements include the development of Partick as an important interchange between rail, subway and bus services and the introduction of direct services between Helensburgh and Edinburgh, the lack of a through city-centre link across the Clyde remains, despite much support for some version of the aspirational *Crossrail* initiative. Glasgow is not alone in being bisected by a river but it could be argued that failure to overcome such a fundamental divide continues to impact on the attitudes of its citizens, whether north or south of the Clyde.

The plan itself has been prepared on a standard Bartholomew depiction, again reflecting the firm's domination in cartographic production. Careful comparison of different maps brings to light a fascinating use of colour to emphasise particular features. In this case, the built-up area is given a pink tinge but of greater interest may be the variation in the identification of public open spaces, cemeteries and gardens between it and, for example, the plan of the city parks of only five years later (**1900**).

1897

The work of the City Improvement Trust

This small and rather unremarkable plan is another example of a document with an important story behind what is shown. It bears testimony to the work of Alexander Beith McDonald as City Engineer and Surveyor and to that of the City Improvement Department. What it cannot show is the existing tensions, when it was produced, between provision of decent homes for the poorest citizens, social responsibility and economic necessity. Considerably diverse opinions existed as to how far the municipality itself should engage in house building and in balancing the demands of amenity, appearance and return on expenditure.

The 1861 census recorded that Glasgow's population had reached 395,503 and, by then, the city was already beginning to experience the side-effects of a rapid increase in residents. Population growth outstripped any rise in housing stock and there was found to be a close association between high density and mortality levels. The squalid conditions of the overcrowded and insanitary closes in the old town have been well documented but it was more the threat of the spread of infectious disease which was the initial spur to action. This was a period of individual involvement, often motivated by a spirit of Christian philanthropy, in seeking to improve conditions for the less fortunate in the community. However, a piecemeal approach would never be effective in dealing with problems of such dimensions. Earlier in his career, Lord Provost John Blackie (1863–66) had helped establish lodging houses expressly for homeless single men and he was a driving force behind a more vigorous and effective campaign.

With the passing of the Glasgow Improvement Act in 1866, the city gained powers to establish a Trust which could purchase and demolish slum properties. This was one of several pieces of legislation behind the long involvement of the administration in services which included the provision of better housing. Initially, the Trust had authority to acquire 36

John Bartholomew, *Glasgow Corporation Improvements* (1897)

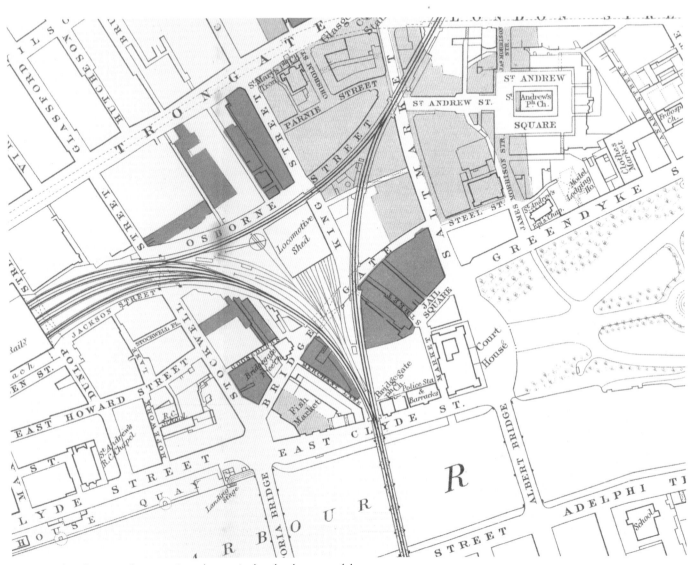

The plan identifies not only properties to be acquired under the terms of the act
but also the location of model lodging houses, as seen off Greendyke Street

hectares of the most congested central areas, concentrated largely around the streets running off Glasgow Cross. In the Gorbals, the ageing cottages of the original village were replaced by modern tenements. Elsewhere, however, progress was slow and the financial crisis which resulted following the spectacular collapse of the City of Glasgow Bank in 1878 stopped both demolition and building in general. By 1891

much of the Trust's land consisted mainly of empty building sites, while private developers, objecting to municipal intrusion into the housing market, held back from building on the cleared areas. This led eventually to the Trust moving into construction itself, as an object lesson, beginning with tenements in Saltmarket. Commencing in about 1888, such activity was to change the face of what was left of Glasgow's medieval core, removing much of the original street pattern and many of its historic houses. It also resulted in the Corporation becoming the largest landlord in the city and reflected a new, more radical faction within local politics.

Under the leadership of Samuel Chisholm in particular, the Trust's focus turned to provision for the poorer citizens. A further Act in 1897 provided powers to purchase land in another seven of the most crowded parts of the old town but, despite this legislation, the basic problem in the whole process was a rather indifferent record in building new houses. Demolition only added to the difficulties faced by those living in these areas, since those displaced were rarely re-housed immediately and some land was also used for laying out new streets. Rentals from city business properties often provided a better return than those from low-income workers' housing. Suburban relocation was a preferred option but the additional purchase of land at Oatlands and Overnewton did little to help, as rents were often beyond the means of the poorest. Nonetheless, these were the first forays into a suburban programme which was to become a major element in the city's housing policy in the twentieth century.

In its time, the city's improvement scheme was the largest and most thorough undertaking of its kind in the nineteenth century. The figures for numbers of houses and shops built can be confusing, largely as the work of the Improvement Trust was only part of the solution. Other bodies were also involved, including the Glasgow Workmen's Dwelling Company, which was more effective at dealing with housing for the poorest workers, and the Statute Labour Department,

responsible for the construction of the impressive St George's Mansions. By 1902, more than 1,600 houses had been built, within thirty-four tenement developments.

The plan, dated May 1897, predates the passing of the Glasgow Corporation (Improvements and General Powers) Act of the same year and is most likely to be related to the bill's submission. The Bartholomew archive records that only fifty-six copies were prepared. It is markedly more detailed than many other contemporary maps produced by Bartholomew and indicates a number of model lodging houses, such as those on Greendyke Street, Drygate and Clyde Street. The other major element of change in the city centre, that of the intrusion of railways, is also clearly seen. However, what is particularly relevant is its indication of those seven areas of land targeted in the Act for renovation, namely the Bell o' the Brae at the junction of Duke Street, High Street and George Street, Goosedubs, Jail Square, King Street, Nelson Street and, south of the river, on Adelphi Street. These are all identified specifically as lands to be acquired under the terms of the bill, while a variety of other areas are shown as scheduled under the Improvement Act of 1866, including the re-designed Gorbals Cross. The sheer extent of the acreage involved underlines the scale of the administration's involvement in rehabilitation. Following on from the work of John Carrick, McDonald was responsible for the preparatory work for this new phase of activity. While appointed as City Engineer, he developed a design skill of his own and was responsible for many civic buildings, including the People's Palace.

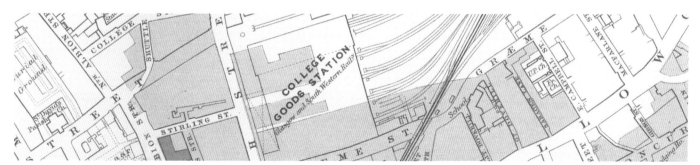

Lands acquired in an earlier act of 1866 are also coloured, as marked on the beside the former College Grounds on the High Street

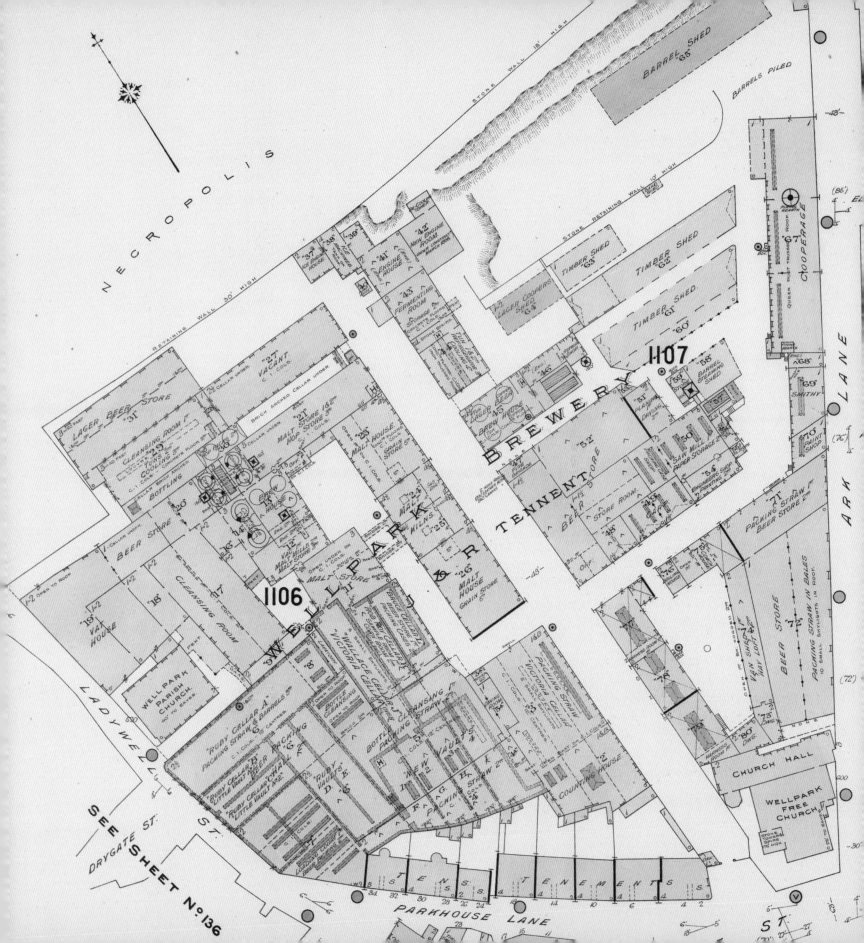

1898

Fire insurance plans: the work of Charles Edward Goad

Increasingly, today's cities have little in the way of heavy industry in their central areas. High rateable values and the impact of environmental health control have tended to result in manufacturing moving out to the more accessible outskirts. One of the striking features of the Victorian urban landscape was the considerable concentrations of industry right at the heart of the city itself. Combined with the high density of buildings and the crowded nature of the older parts of towns, the presence of activities which dealt with frequently dangerous, if not combustible, materials made the outbreak of serious conflagrations more of a possibility. The commercial interests of business led to the development of a specialised type of map, which is generally described as a fire insurance plan. These maps provide information to meet the under-writer's requirement to know the structure and layout of buildings, along with the location of policy holders.

While there are several surviving plans of parts of London

specifically produced for insurance offices and dating from the first half of the eighteenth century, the extensive production of this style of cartography really began to develop only after the foundation of the Sanborn Map Company in New York in 1867 by Daniel Alfred Sanborn (1827–83). Only two years later a young Englishman, Charles Edward Goad (1848–1910), emigrated to Canada, where he was engaged in work on a number of railways. While employed as an engineer, he began preparing insurance surveys of Canadian cities and there can be little doubt that he was influenced by the work being conducted by Sanborn, since he eventually bought out the company's Canadian interests. Given the nature of the building materials at that time, fire was a significant risk and, in 1875, the Charles E. Goad Company was established in Montreal to extend production of these plans.

During the next decades, his teams of surveyors mapped over 1,300 Canadian towns and in 1885 Goad returned to

Charles E. Goad, *Fire Insurance Plan of Glasgow, vol.4, sheet 138* (1898)

England to open a London office. The company rapidly established a dominance in this cartographic field and extended its operations to Europe, Egypt, the West Indies, South America and South Africa. Before the end of the century, all the major British towns had been mapped. For Scotland, centres mapped included Campbeltown, Dundee, Edinburgh, Glasgow, Granton, Greenock, Leith and Paisley. As this list suggests, the focus was on industrial and commercial districts, as well as ports. The first plans of Glasgow itself are dated 1887, and by 1901 five atlas volumes had been produced. While the earliest sheets mapped properties in the city centre, this was taken to be an area west of the High Street and south of Cathedral Street. Later atlases covered business properties south of the river, in Garnethill, Anderston and Cowcaddens, in Calton and Bridgeton and the dock area to Govan and Partick respectively.

Goad's work was significant in standardising both survey and reproduction processes, for he appears to have settled early on an accepted choice of colouring to indicate building materials: pink for brick, light brown for brick and timber, yellow for timber and blue for glass. The plans were lithographed, grouped by district and preceded by an index sheet, with a standard key explaining the various signs used. Usually drawn at a scale of 1:480, they display an impressive level of detail on building use, ownership, construction material, height, number of floors, property lines, special fire hazards (such as chemicals or ovens), presence of windows and water supplies. Although they are not well known or widely available, they provide unique source material, additional to the Ordnance Survey, about building types, land use and urban layout for a period of remarkable change in most British cities. Unfortunately, their regular revision, after about five years or so, often meant that earlier versions were either destroyed or pasted over. The Goad business was to continue as an independent company until 1974 but the name still survives, as a trademark, as part of the Experian Group.

The first selected illustration displays the Tennent's Wellpark Brewery just off Duke Street. Prepared in 1898, it provides a valuable example of the wealth of information to be gleaned from this type of survey, with such features as the indication of the varying uses on different floors of business premises and buildings under construction. Once again, it emphasises the close juxtaposition of industry and homes, listed as tenements (TENS). In the January of that year, four firemen died in an explosion in a chemical works on Renfield Street.

While Glasgow had experienced devastating fires in the seventeenth century, it was in the post-war period of the 1960s that it gained an unenviable reputation as the 'Tinderbox City' after a number of major incidents. This second map detail is a reminder of the tragedy behind the map. It shows the bonded warehouses owned by Arbuckle Smith & Company at number 83 Cheapside Street, Anderston where, on 28th March 1960, a fire broke out which resulted in one of Britain's worst peacetime fire disasters. The building contained over a million gallons of whisky, as well as 30,000 gallons of rum. As the temperature of the fire increased, a massive explosion blew the front and rear walls out, killing fourteen firemen and five men from the Glasgow Salvage Corps engaged in trying to control the blaze. This was to be followed eight years later by an equally horrific loss of life when 22 employees died in a factory fire in nearby James Watt Street. The Goad plans underline the difficulties faced by the fire and rescue services of the time, for many industrial premises were still located in cramped and narrow streets.

OPPOSITE. This extract provides a detailed depiction from the Goad sheet 346 (1943) of the Arbuckle Smith property in Cheapside Street

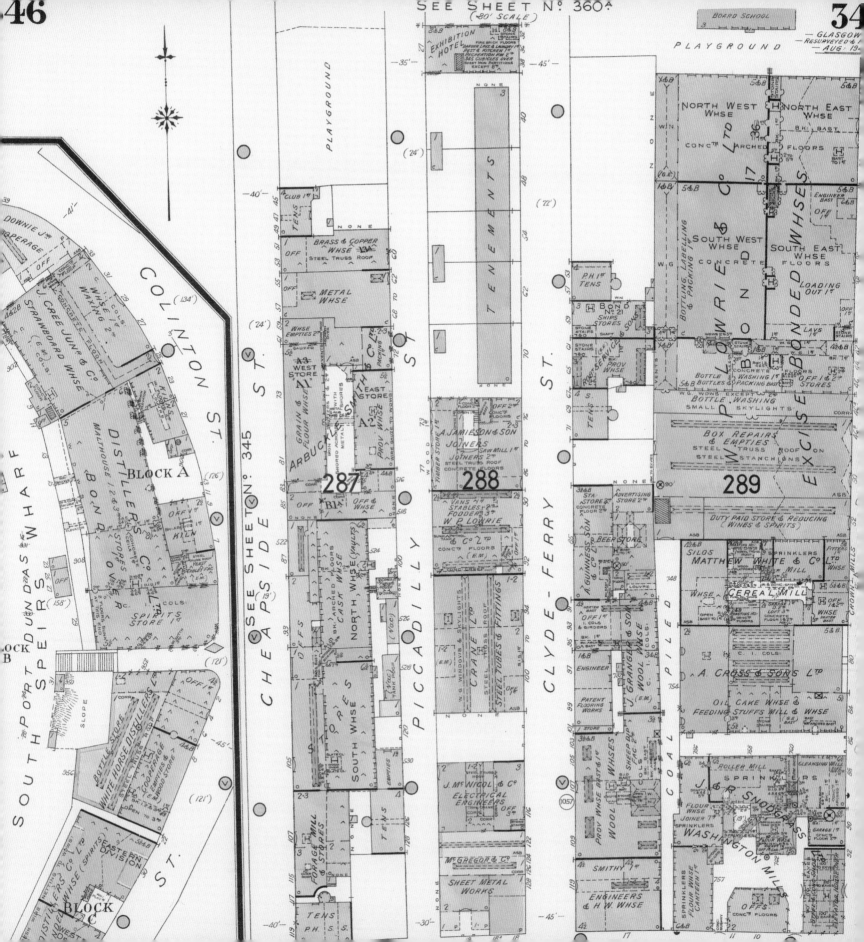

ELEVATION OF PROPOSED ERECTION SHEWING ITS HEIGHT IN CONTRAST WITH THE HEIGHT OF THE HOUSES FRONTING THE RIVER.

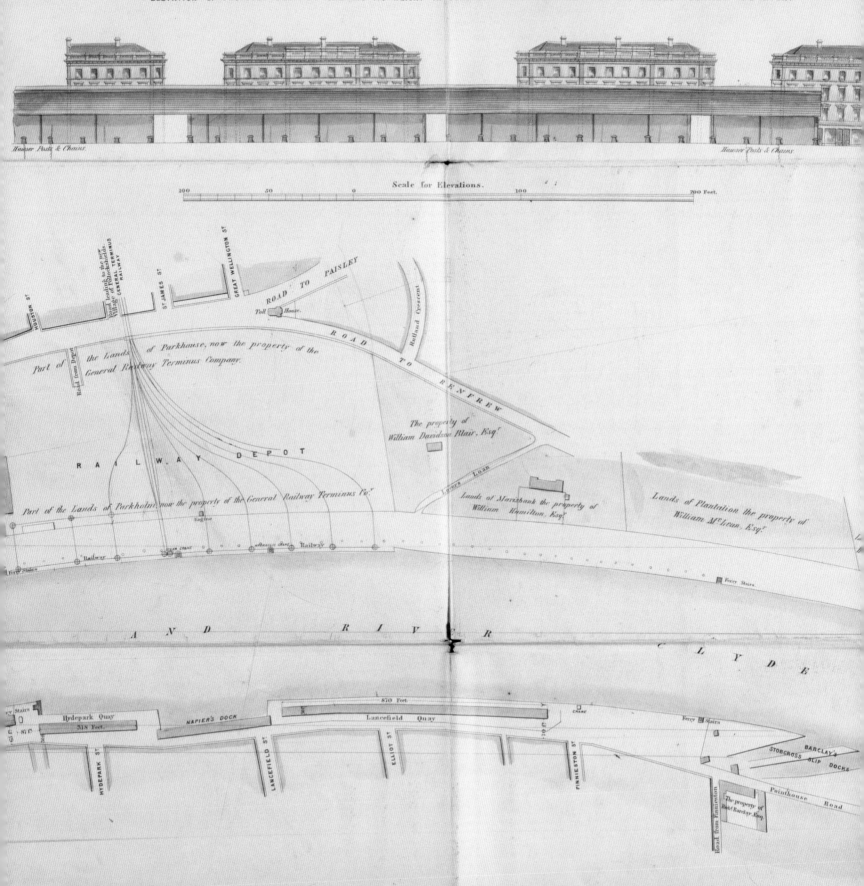

1898-99

Glasgow's harbour develops

No narrative of Glasgow's growth could be complete without due recognition of the place that the River Clyde holds in the story. Reference has been made already to the difficulties in navigating the river up to Glasgow, to the success of a series of engineers in overcoming these problems and to the association with shipbuilding which made the term 'Clyde built' a guarantee of craftsmanship and quality. The close affinity between the two has changed immeasurably over the last two centuries but it is certainly true that it was the expertise and determination of its civic leaders which converted what was a restricted, shoal-filled channel into a waterway route to the markets of the world for both Glasgow's citizens and its manufacturers.

By 1864, artificial deepening of the river by a process of virtual canalisation in certain stretches, combined with regular dredging, created a minimum low water depth of 12 feet upstream to the city. This not only allowed the largest trading ships to reach Glasgow but also enabled local shipbuilders to launch and fit out vessels which could then reach the deeper waters of the estuary. Nature worked against navigation in another way. The prevailing westerly winds could create a packed harbour in the days before steam tugs came into general use, leading to the construction of the towing path on the south bank which is a feature of several of the nineteenth-century maps of the city.

The advent of steam power was to be a major factor in transforming the port facilities at Glasgow but access for vessels was only one aspect of harbour development. Quays, warehouses, storage sheds, docks and their attendant facilities, combined with ready access to land transport for onward trans-shipment, were equally important to a proper infrastructure for growth. In Glasgow's case, the extensive railway layout at the General Terminus Quay was vital for bulk goods handling. This first image is taken from an 1852 improvement

Black & Salmon (architects), *Plan of the Improved Harbour of Glasgow* (1852)

plan by Black & Salmon, a local firm of architects. It shows elevations of proposed bonded storage sheds and wharves on the north bank at Parkholm but also indicates that early efforts focussed mainly on extending quays along the water's edge. Between 1801 and 1881, quay length grew from 349 metres to nearly 7.7 kilometres.

Although initial rebuilding of the Broomielaw, combined with the addition of Steamboat and Anderston Quays, were signs of intent, congestion in the harbour was commonplace. In 1858, the Clyde Navigation Trust took over responsibility for the management of the river and its trade. Significantly, this body comprised shipbuilders, merchants and industrialists working together. Nonetheless, it took several years of pressure to tackle the problem of sufficient berth space and it only began to be resolved with the construction of Kingston Dock at Windmillcroft in 1866. This tidal basin had been marked on a series of maps as a proposed dock since the late 1830s. It was the first such facility built outside, and parallel to, the main channel and was to be followed in 1880 by the larger Queen's Dock at Stobcross. Growth in both trade and the tonnage of shipping soon filled the accommodation in these berths. A third solution was eventually completed with the opening of the three parallel basins constituting Prince's Dock at Cessnock between 1893 and 1897.

While the channel to Glasgow was narrow, downriver it opened into a wide and relatively safe estuary. Much of the growth in both the city and its harbour was related to the enormous expansion of coastal trade during the early nineteenth century. Increasingly, steamers connected Glasgow with a range of ports in Lancashire, Cumberland, Ireland and the west coast of Scotland, strengthening its role as a regional distribution centre. Links to remote communities along the western seaboard survived competition from other means of transport well into the twentieth century and those workhorses of such commerce, the Clyde puffers, were to be immortalised in the short stories of Neil Munro published in the *Glasgow Evening News*.

A major factor in harbour congestion was the rapid growth in steamers carrying passengers to a wide variety of Clyde coastal resorts. A visit to the Riverside Museum at Pointhouse Quay provides ample opportunity to appreciate the variety and type of vessels involved in this trade. As the harbour grew, so commerce became increasingly foreign and, particularly, imperial but the expanding shipbuilding industry, which was a key element in Glasgow's economic growth, created difficulties in the smooth operation of the port, largely because the shipyards were interspersed with the docks. A spatial separation of vessels and goods only fully developed in the twentieth century.

This detailed plan was produced by James Deas soon after

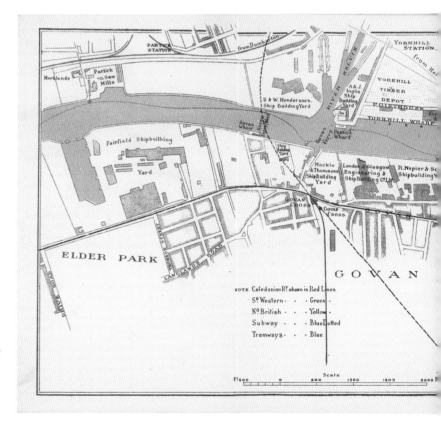

the completion of Prince's Dock. Deas served the Clyde Navigation Trust as engineer-in-chief for thirty years. Prior to this, he had worked in the offices of John Miller, the Edinburgh railway engineer, and himself later became chief engineer of the Edinburgh and Glasgow Railway. Considered an eminent authority, it was under his direction that many major additions and the three graving docks at Govan were constructed. A prolific writer, Deas also presented several papers on the river to a range of engineering bodies. Tragically, he died at the end of 1899, while lunching in a Glasgow restaurant.

Surviving documents show that this plan is, in fact, a portable colour reduction of the Bartholomew original produced by Maclure, Macdonald & Company but with the added advantage of an accompanying guide, which lists the owners and shipping agents involved, as well as the particular trade the vessels were employed in. Some element of separation had developed already, with berths in the upper and lower harbour being retained for coastal shipping. More than 25 companies are identified, including such names as MacBrayne and Burrell & Son.

James Deas, *Hand Map of Glasgow Harbour and Docks* (1899)

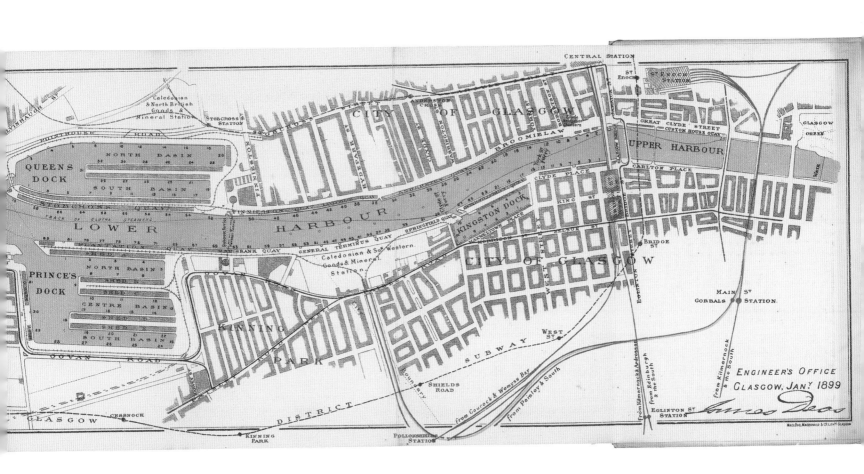

1900

The dear green place

Several theories exist about the origin of the name Glasgow but the leading contender is that it derives from the Brythonic Celtic words 'glas', meaning green, and 'cau', meaning hollow or 'place of the green hollow'. This has also been interpreted as 'dear green place'. Brythonic was the language of those inhabiting Strathclyde in the sixth century AD. Despite the reputation that the city has had as an industrial powerhouse, this map celebrates the fame it also enjoys for the extent of its parks, gardens and green places.

It also recognises the contribution of two leading public servants, namely Alexander Beith McDonald (1847–1915) and James Whitton (1850–1925). While McDonald is recorded as City Engineer, he was *de facto* City Surveyor from 1890 to 1914 and was responsible for the layout of Bellahouston Park on its purchase in 1895–96. Although a native of Stirling, he was articled to the Glasgow surveyors Smith and Wharrie in 1862, before entering the City Architect's office eight years later. While there, he assisted in work relating to the City Improvement Trust and was to play an influential role in this area throughout his career.

Whitton was Superintendent of Glasgow Parks. In 1893, he succeeded Duncan McLellan, who had served in this post for forty years. The following year, McLellan published *Glasgow Public Parks*, which describes in some detail all the green areas shown on this map. In 1878, he had taken an extended tour of European cities, growing to appreciate the benefits to urban dwellers of fresh air and open spaces. This was part of a conscious effort by the authorities to improve the well-being of its citizens by learning from experiences elsewhere. To underline this commitment, it should be noted that in the last years of the century the Corporation was spending over £37,000 annually on park maintenance alone.

John Bartholomew, *Plan of the City of Glasgow Shewing Parks, Gardens and recreation Grounds belonging to the Corporation of Glasgow* (1900)

Whitton was also dispatched abroad in 1897 where, particularly in Germany, he was greatly influenced by the ideas of creating open spaces for the appreciation of nearby surroundings. Subsequently, he became a major figure in the development of urban parks, at a time when there was a growing awareness within many local authorities of their benefit to citizens.

The plan indicates those areas directly under Council control and, therefore, hints at their wide variety. Ranging from Bellahouston's 72 hectares to the diminutive 1.6 hectares of the Govanhill recreation ground, and from the more open lands of Cathkin to the formality of Maxwell Park, they appear as a chain of green pockets circling the city. The oldest is the most central. King James II granted the lands now known as Glasgow Green to Bishop Turnbull and the people of Glasgow in 1450, thereby establishing its first tract of public open ground. The present park is made up of what were once distinct areas, such as the Gallowgate Green and Flesher's Haugh. However, it was only when James Cleland was Superintendent of Public Works that it became something more akin to a park. Between 1817 and 1826, he put unemployed weavers to work laying paths, planting and shaping the lands into a space more fit for recreational use.

The Green has played an integral part in the city's story, fulfilling many roles – agricultural grazing land, a communal washing and drying green, as well as the site of fairs, public executions and live concerts. In more recent times, the Green has seen massed rallies in support of the Reform Act and women's suffrage, in addition to demonstrations against war in its many guises, such as that addressed in 1914 by John Maclean, the leading Scottish socialist, and, by a curious twist of irony, the 2003 protest against the Labour government's support of the invasion of Iraq.

While Glasgow Green is tied closely to the city's early history, other parks are more a product of the Victorian era. Kelvingrove in particular is redolent of the period. Laid out in 1852 by Sir Joseph Paxton, the leading English gardener of the day, it was the first purpose-designed public park in Scotland. It covers more than 30 hectares and the fluid layout of its paths reflects one of its main features, the River Kelvin. The map identifies, standing above the river, its most impressive structure – Kelvingrove Museum and Art Gallery, one of Glasgow's most visited and popular buildings, built partly on the profits from the 1888 Exhibition.

Equally Victorian in style are the Botanic Gardens which moved to their present, more extensive site on Great Western Road in 1839–42. They remained a private institution until responsibility passed to the Corporation in 1891. Queen's Park (1857), another Paxton design, provides a similarly open space for the city's south side and is named after Mary, Queen of Scots, whose army was defeated at the nearby battle of Langside, again marked on the map. The largest open area in east Glasgow, Alexandra Park, was bought in 1866 by the Improvement Trustees and was created from agricultural land by several hundred unemployed labourers during what was another trade depression. Much of the early planting was affected by industrial pollution, underlining the important point that open space alone is not enough to improve an environment. Elsewhere, Springburn (1892), Ruchill (1892) and Cathkin Parks (1887) all testify to the growing acquisition or donation of land for public recreation. Observant readers will notice that some parks lie outside the boundary. The plan also shows several graveyards and other small parks under the city's care.

As the records indicate that only 120 copies of the plan were printed, it is unlikely that it was intended for general circulation. Today, the city's parks continue to play an equally important role in providing recreation, sport and entertainment for residents and visitors alike. Council investment in refurbishment and replanting paid dividends when the city hosted the XX Commonwealth Games in 2014 and, as important sports venues, the parks played an integral part in promoting Glasgow.

OPPOSITE. The whole map gives a good impression of a city surrounded by sizeable pockets of green space

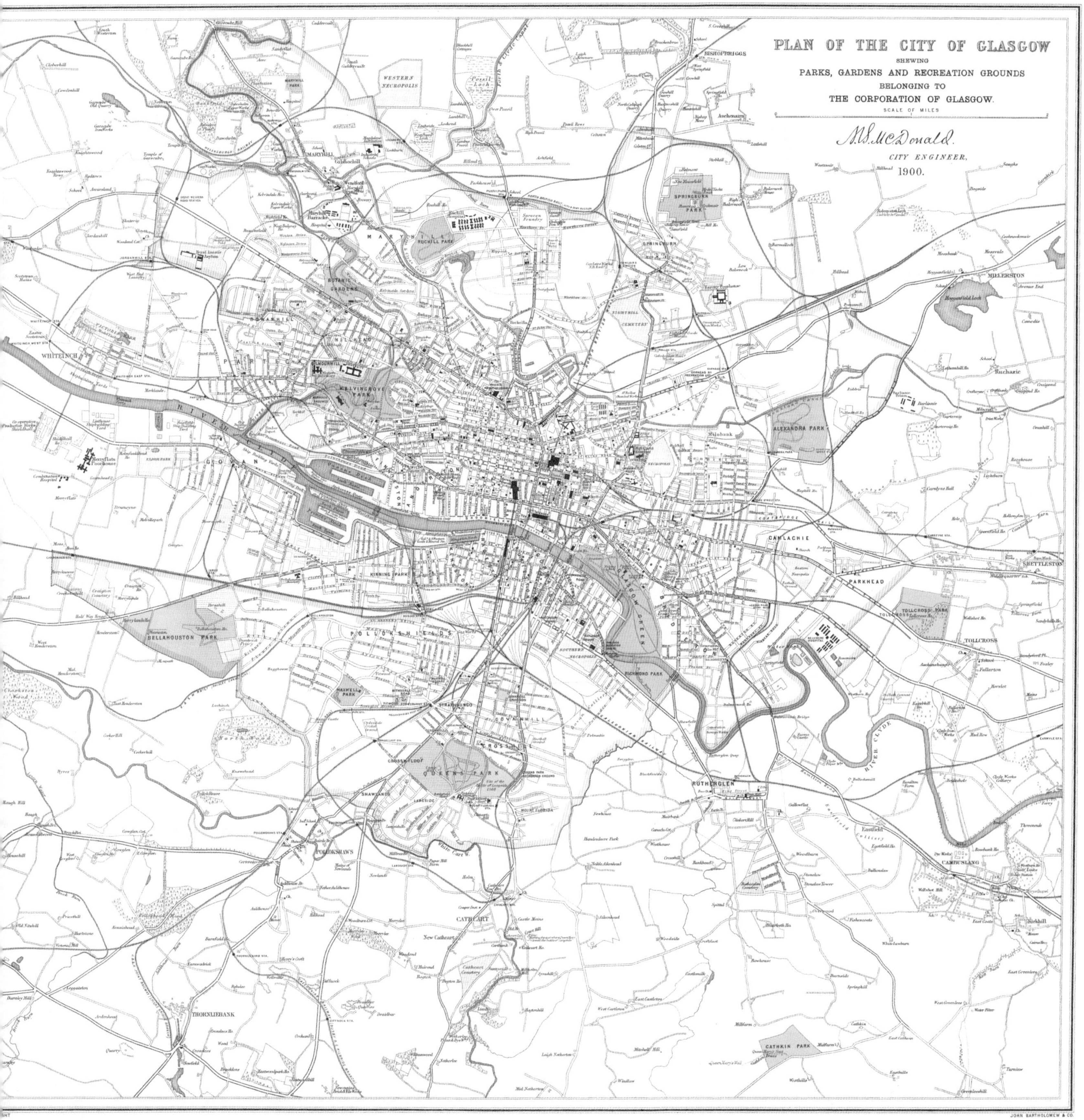

PLAN OF THE CITY OF GLASGOW

SHEWING
PARKS, GARDENS AND RECREATION GROUNDS
BELONGING TO
THE CORPORATION OF GLASGOW.

SCALE OF MILES

A. B. McDonald.

CITY ENGINEER,

1900.

120 Copies

JOHN BARTHOLOMEW & CO.

1901

A *new international exhibition for a new century*

The success of Glasgow's first international exhibition (1888) resulted in a healthy profit as a solid basis of funding for the city's proposed new Art Gallery and Museum. With public contributions more than doubling this sum, the Corporation provided a site at Kelvingrove and, within ten years, the foundation stone was laid in 1897. It was suggested that its opening be celebrated with a second major exhibition, which coincided conveniently with both the jubilee of the Great Exhibition at the Crystal Palace and the beginning of a new century. The assurance and optimism of the civic leaders is clearly visible in this willingness to repeat such a major event, so relatively soon after the 1900 Exposition Universelle held in Paris.

In some respects, the timing was rather unfortunate. The death of Queen Victoria in January 1901 meant that the royal court went into mourning, which prevented the new king, Edward VII, from attending either the opening ceremony or the exhibition itself. It also took place while the controversial Anglo-Boer War was still in progress, adding to Great Britain's diplomatic isolation in Europe, underlined by the lack of any pronounced German presence during the Exhibition. On the other hand, the city's record in engineering, manufacturing, art and design, as seen in the growing significance of the distinctive 'Glasgow Style', gave the authorities the confidence to highlight Glasgow's position as the second city of the British Empire. The additional stated values of promoting peaceful international relations appear today as somewhat more illusory than a reflection of real conviction.

Nonetheless, Glasgow was determined to impress and planning of the whole enterprise was on an ambitious scale. In the end, the event was markedly more international than its predecessor, with Holland, Russia, Sweden, Morocco, Mexico, Spain, Japan and Canada being among the nations with pavilions. Centred on the new Kelvingrove building, which

John Bartholomew, Glasgow Exhibition: route map from the terminal stations and hotels (1901)

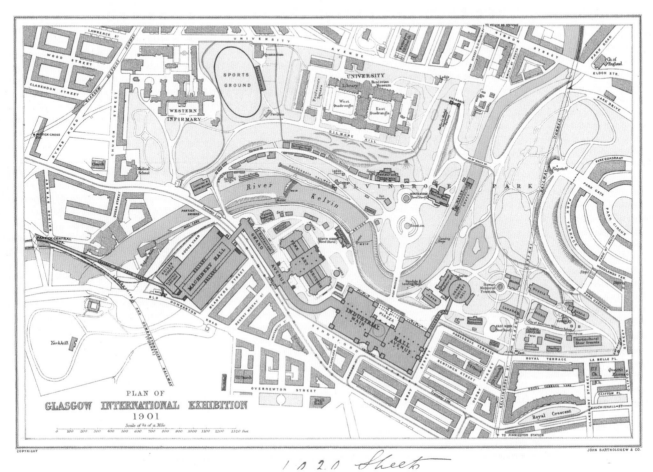

An entirely separate Bartholomew plan of the immediate area of the International Exhibition,
detailing the individual buildings and the covered route from Partick Central Station

housed the fine art displays, it encompassed over 29 hectares and included an industrial and machinery hall (at Bunhouse), model farm, Indian theatre, concert hall, sideshows and various tearooms. The whole layout can be seen on this Bartholomew plan, which includes the covered routes and Grand Avenue designed to protect visitors from the worst of the Glasgow rain. As it turned out, the opening months were blessed by unnaturally good weather. Interestingly, the tramlines and the route from Partick Central Station are clearly identified, whereas the

nearby Partick Cross Subway station is much less obvious, possibly underlining the desirability of a noble approach to the event. Certainly, it was the tramway system which pushed forward its electrification specifically to enable the smooth transport of large number of visitors around the city.

Local commercial enterprises and industries were again to the fore. G. & J. Weir of Cathcart displayed their range of pumps with a waterfall in the Machinery Hall, whereas Singer's sewing machines appealed to a more domestic market.

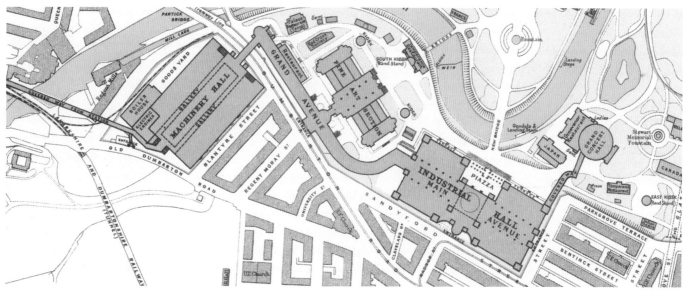

The comparative size of buildings emphasises the importance to Glasgow of displaying its engineering and industry at the Exhibition

Leading department stores, such as Forsyth's, Wylie & Lochhead and Pettigrew & Stephens had display stands, while other household names participating included Fry & Sons, Lipton's and Heinz. By this date, nearly half of the world's shipping was either built or powered by companies based on the Clyde and the Grand Avenue was used to house a popular arrangement of ship models. This was crowned by Fairfield's recently launched HMS *Good Hope*, then the British navy's most powerful armoured cruiser. While a source of pride and critical interest, the vessel was to be the first major naval casualty of the Great War when it sank with all hands at the Battle of Coronel in November 1914. Despite the promotion of local industry, many competitor firms from both home and abroad also made an appearance, though few visitors would recognise the threat to the city's industrial base at this date.

Although a contrast to the standard Scottish housing style, the Sunlight Cottages erected by the soap manufacturers Lever Brothers remain in Kelvingrove Park as a reminder that the living conditions of the less affluent was being addressed by private enterprise as well as municipal authorities. Musical

entertainment was provided by a cross-section of bands, including that of the American 'March King', John Philip Sousa, and Dame Nellie Melba gave three concerts. As the visiting public's expectations became more demanding, there were many other features, such as a switchback railway and water chute, as diversions from the more commercial aspects of the exhibition itself. Sports events were organised for the new University stadium and the city played host to several conferences that year, including that of the British Society for the Advancement of Science. One notable difference from the 1888 event was the novel presence of automobiles which were not only on display but took part in competitive trials.

The second plan, from the same year, is a striking depiction of the Glasgow District Subway loop and its fifteen stations prominently dominating a backdrop showing the city, its parks and the Exhibition Grounds. Construction on it began in 1891 but services were not fully operational until January 1897. James Miller, the Exhibition architect, provides a link, as he also designed the St Enoch station, which originally housed the Subway headquarters. Called many things in its time – 'Victorian toy', 'The Shoogle' or 'Clockwork Orange' – the Subway holds a special place in many a Glaswegian heart and is an integral part of the city's transport infrastructure. Presently in a process of modernisation, in 2013–14 it carried more than 12.7 million passengers. Although it is the third oldest system in the world, it is also one of the smallest and has never been extended.

By the close of the Exhibition in November, nearly 11.5 million visitors had attended and a profit of £39,000 had been made, facilitating the purchase of additional works of art for the new galleries. Glasgow was to further benefit when Sir William Burrell, who had served as an Exhibition committee member, subsequently gifted his considerable art collection to the city in 1944.

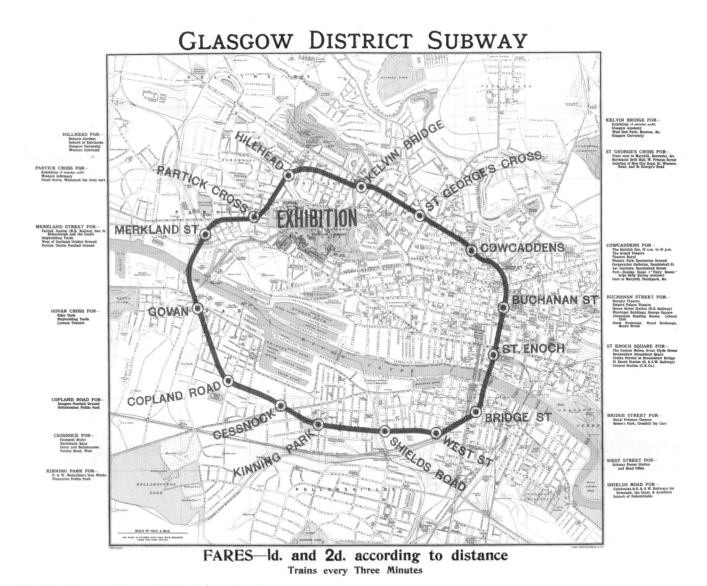

John Bartholomew, *Glasgow District Subway* (1901)

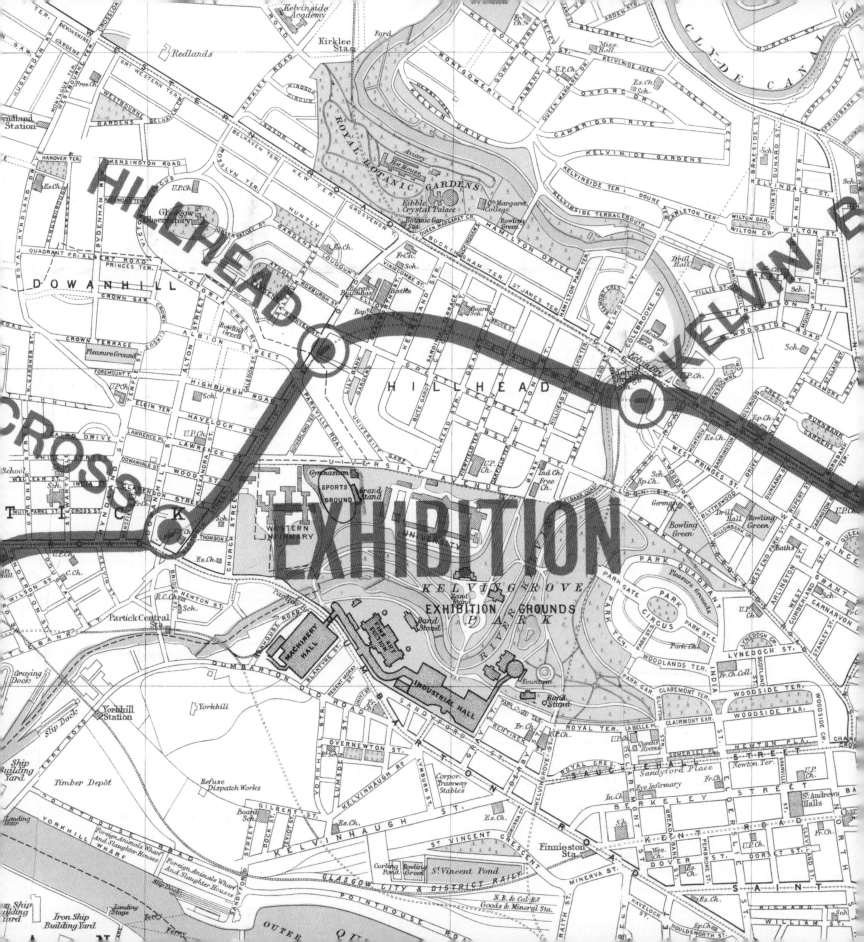

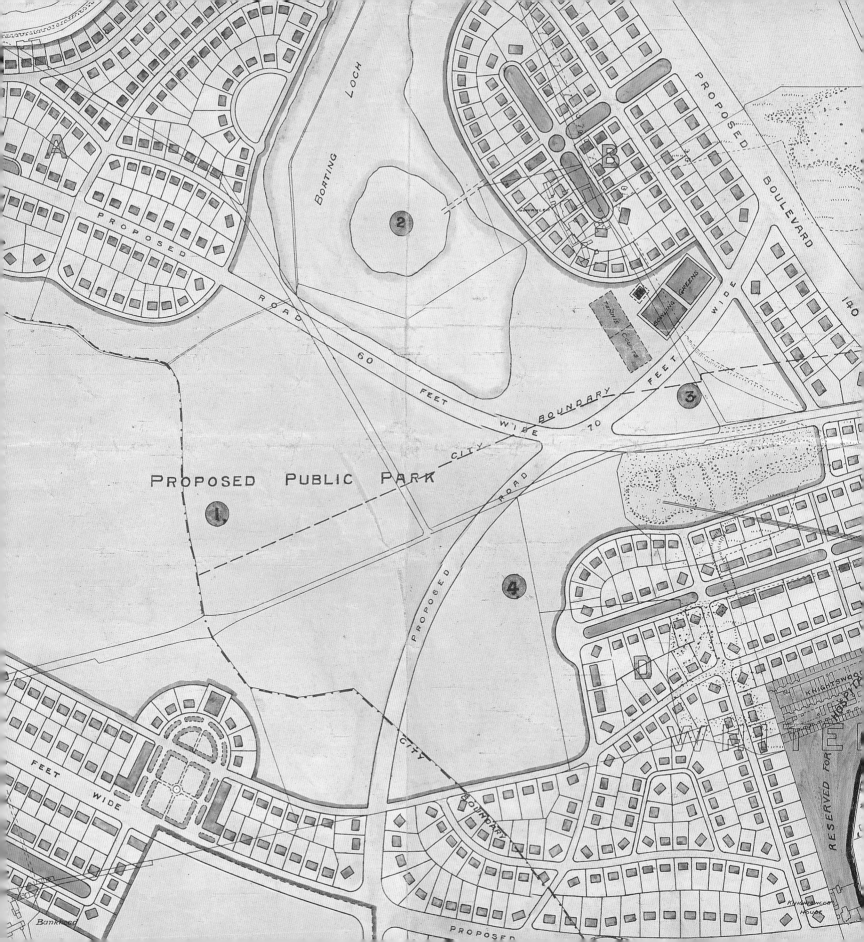

1921

Knightswood: *a new housing scheme for a changing city*

The inter-war period of the 1920s and 1930s saw a major transformation in the appearance of Glasgow. By 1910 the Scottish economy was slowing down and the years of the First World War saw little in the way of house construction. A Royal Commission in 1917 into housing conditions and a subsequent government report stressed the need for major improvements, particularly for Glasgow and other urban areas in the Clyde Valley. The expectation that many of those who had fought in the trenches would return to 'homes fit for heroes' added to the pressure for reform. It has been calculated that, in 1921, 63.5 per cent of the city's population lived in one- or two-roomed accommodation and that more than 55 per cent of those dwelt in what were deemed overcrowded homes. As the economic depression of the 1920s grew, the state was increasingly drawn into attempts to improve the situation.

Post-war legislation, in particular the Addison Act, named after the then Minister of Health, introduced a somewhat tentative approach to town planning and state subsidies for building, which led directly to the creation of a municipal housing department in 1919. The scale of the housing problem required a radical approach which was met by the acquisition of large areas of suburban land and a degree of uniformity in construction design. Between 1926 and 1938, Glasgow more than doubled in size but, regardless of the difficulties faced, the city managed to build more than 54,000 houses in the period up to 1944, mostly within a fifteen-year time span. This Knightswood plan shows how rural farmland was altered into a more regular suburbia, with the only concessions being the fluid sweep of the road pattern. Such a layout is characterised by broad avenues, in particular the proposed Boulevard extension of Great Western Road, constructed between 1924 and 1927 with a generous width of 140 feet. It shows clearly the landscape underlying the proposals in such features as

John Bryce, *Proposed Housing and Park Scheme at Knightswood, Bankhead &c* (1921)

233

Blairdardie brick and tile works, Bankhead farm and the lines of industrial tramways. Prepared before harsh economy and political expediency resulted in a poorer standard of construction, it reflects the influence of Ebenezer Howard's belief in the garden city.

Glasgow Corporation purchased the farmland of Knightswood from the Summerlee Iron Company in 1921 and this original design by John Bryce for the City Engineer was approved that year. Bryce was to become Director of Housing two years later, about the time when the first phase of construction began. Initially, fifteen different types of two-storey, cottage style semi-detached houses and flats were built, often with Arts and Crafts elements incorporated into the design. These were also introduced in Riddrie and Mosspark and are in marked contrast to the extensive areas occupied by private sector bungalows in other suburban areas. Knightswood stands out as the most extensive development not only in Glasgow but in the whole of Scotland. Dated 1921, this plan was prepared when a considerable part of the proposed development lay outside the city. It was only after the 1926 boundary extension that the later phases were inaugurated.

Overall, four separate areas are indicated south and west of Great Western Road where more than 1,700 houses would be built. Costs were cut drastically for later phases, particularly after 1931, when subsidies were withdrawn. Eventually, more than 6,000 houses were constructed to accommodate 25,000 to 30,000 people. At this scale, it was a size similar to the satellites proposed by the Garden City Movement, which advocated the provision of associated social, recreational and commercial facilities. Knightswood was built as a residential suburb and, for several years, few facilities were provided for those citizens re-located to this more distant part of Glasgow. The first school and church were not opened until 1932 and, as can be seen from the plan itself, there was little in the way of shops or opportunities for employment in the immediate area. Although demand might, in time, create facilities, this was not the same as allowing for these as part of the initial design concept. This contrasts with the planned new townships of Rosyth in Fife or Gretna in Dumfriesshire and indicates how far planners had strayed from the original Garden City vision.

The layout was a 'child of its time' and is noticeable for having no public houses but a considerable area devoted to public parks, tennis courts, bowling greens and open spaces, with some land reserved for an extension to Knightswood Hospital. In an era when few people owned cars, most residents had to rely on public transport to commute to work and the nearest local railway station (Scotstounhill) was inconveniently situated at the edge of the plan. As travel costs mounted, the residents' sense of isolation increased. This was an experience that was to be repeated in later and more remote developments in Drumchapel and Easterhouse.

Immediately after the end of the Second World War, housing in the form of mass-produced prefabricated homes was erected in several parts of Glasgow, including Knightswood, under the Temporary Housing Programme, to provide homes for the many families in need of accommodation. The 'prefabs' proved to be surprisingly popular with their occupants and many lasted far longer than their allotted lifespan of ten years. These were subsequently replaced by high-rise flats in the 1960s. Today, Knightswood's low density of housing makes it one of the city's more desirable residential areas.

OPPOSITE. Bryce's plan overlays an existing landscape with a design based around a proposed central public park with an extension of Great Western Road (proposed boulevard) as an initial boundary

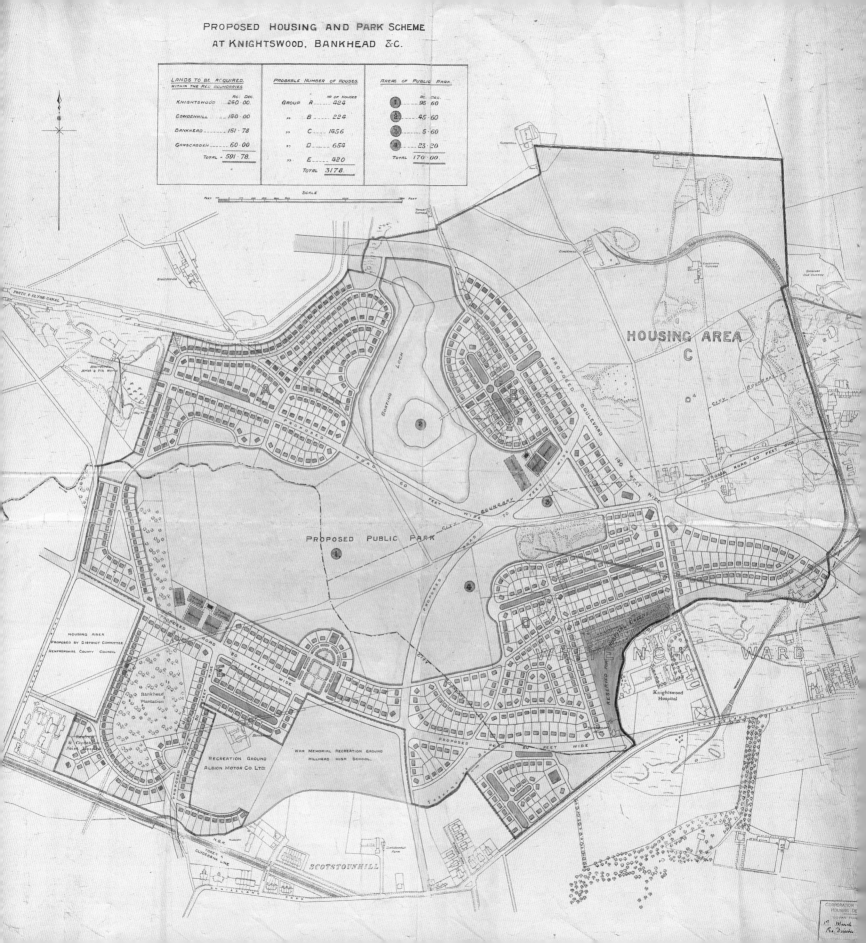

PROPOSED HOUSING AND PARK SCHEME
AT KNIGHTSWOOD, BANKHEAD &C.

LANDS TO BE ACQUIRED, WITHIN THE RED BOUNDARIES		PROBABLE NUMBER OF HOUSES		AREAS OF PUBLIC PARK	
	AC: DEC.		N° OF HOUSES		AC. DEC.
Knightswood	240·00	GROUP A	424	1	95·60
Cowdenhill	140·00	" B	224	2	45·60
Bankhead	151·78	" C	1456	3	5·60
Garscadden	60·00	" D	654	4	23·20
Total	591·78	" E	420	Total	170·00
		Total	3178		

SCALE

PROPOSED PUBLIC PARK

HOUSING AREA C

BOATING LOCH

PROPOSED BOULEVARD

HOUSING AREA
Proposed by District Committee
Renfrewshire County Council

Bankhead Plantation

WHITEINCH WARD

Knightswood Hospital

RECREATION GROUND
ALBION MOTOR CO LTD

WAR MEMORIAL RECREATION GROUND
Hillhead High School

FORTH & CLYDE CANAL

SCOTSTOUNHILL

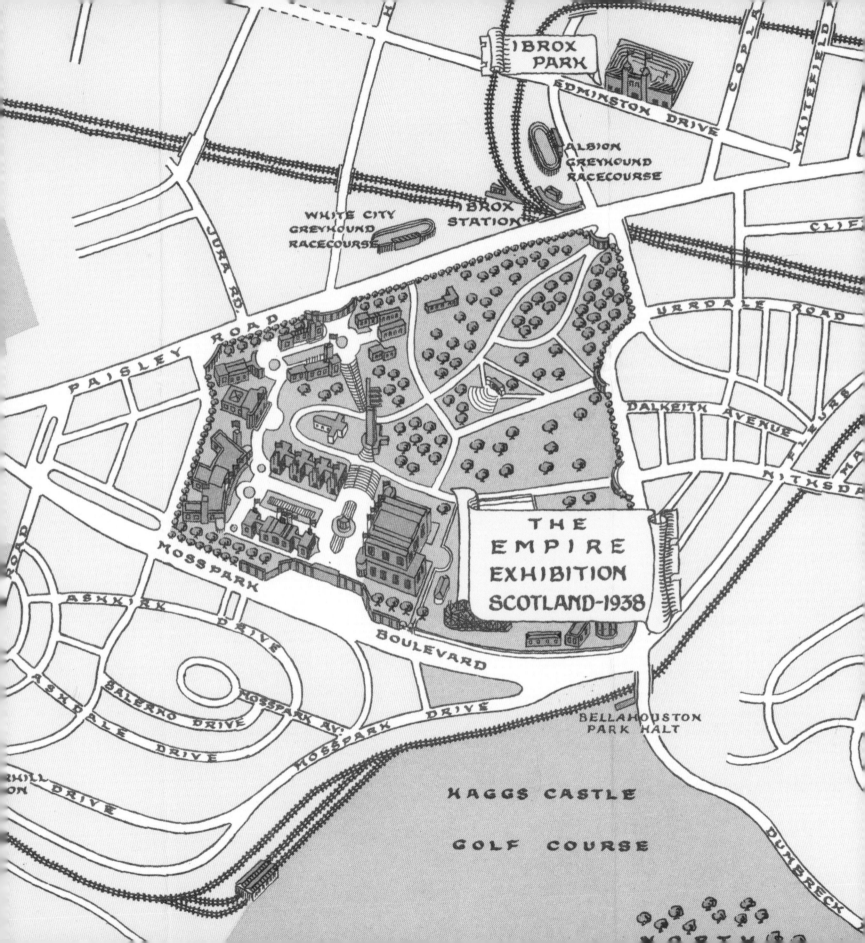

1938

The Empire Exhibition

Fifty years after its first International Exhibition, Glasgow was to host the Empire Exhibition in 1938, at a time markedly different from that of its predecessors. The west of Scotland was only slowly recovering from the deepest economic recession of the century. Between 1928 and 1939, unemployment doubled, and in the same period the total labour force in Lanarkshire contracted by 55 per cent. As orders dried up, output in both shipbuilding and steelmaking dropped dramatically. Glasgow's economy was still overly reliant on heavy industry and was hit harder than many other parts of the country.

In other ways, it was an inauspicious time to hold such an international celebration of empire. The worsening European political situation was a backdrop to the festivities. Before the Exhibition closed, the British Prime Minister Neville Chamberlain flew to Munich to meet Adolf Hitler and signed an agreement which was to delay hostilities for a year. In addition, one of the major countries of the British Empire, India, did not participate in Glasgow, a sign of change within this 'family of nations'. What the organisers could never have predicted was that the event would take place in dreadful weather conditions, during one of the wettest and coldest summers in years. Despite this, more than 12.5 million visitors attended over the six months that it was open.

As 'Second City of the Empire', Glasgow had excellent credentials for hosting such an event. It was a premier port, as well as a major manufacturing and mercantile centre. Furthermore, it had excellent transport links, with more than half of Scotland's population living within 25 miles of the city. Significantly, Edinburgh City Council contributed £10,000 to the organisational funds. In contrast to previous exhibitions, the site chosen for it was Bellahouston Park, as it had sufficient level ground. Although the objectives of the organisers were to stimulate industry and trade, the intention was also to

John Bartholomew, *Pictorial Plan of Glasgow* (1938)

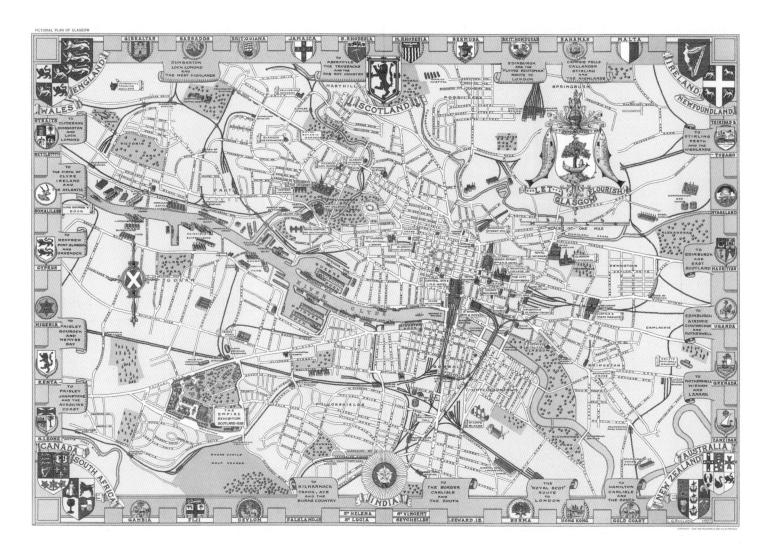

instruct and entertain. Despite incurring a financial loss, many considered that it brought joy, wonder and a sense of hope to those who visited that summer. For a brief period, Glasgow was the centre of attention at a time when its political power and industry were on the wane.

Thomas Tait, architect of the recently completed St Andrew's House in Edinburgh, was put in charge of the design of the Exhibition grounds and his team was to include Basil Spence and Jack Coia. He would be remembered for the symbolic Tower of Empire which would dominate the site,

nicknamed 'Tait's Tower' by the Glasgow public. Three distinct categories of pavilion resulted – those of the Dominions or Colonies, those reflecting the life and culture of the British Isles, and buildings demonstrating achievements in engineering and industry. Form and colour were used in stark contrast to the surrounding residential areas, while commentators noticed that the city had achieved something more dignified but less solemn than the previous Empire Exhibition held at Wembley.

With an additional focus on entertainment, an Amusement

Park, organised by Billy Butlin, was laid out beside the main grounds. Its sense of pleasure was arguably better remembered by many of those who attended than the more 'worthy' exhibits. The visitors included several members of the Royal family, after King George VI officially opened the event in May, as well as other popular or distinguished individuals, including John Buchan, Governor-General of Canada and famous novelist, Paul Robeson, the renowned American singer and Donald Bradman with the visiting Australian cricket team. In the Concert Hall, visitors could enjoy music from such leading luminaries as Gracie Fields, the dance bands of Henry Hall and Jack Hylton, and major London orchestras under the batons of Sir Henry Wood and Sir Adrian Boult.

During 1938, several guides to both Glasgow and the event itself were published, frequently accompanied by plans of the city and the Exhibition grounds. None, however, are as striking as this 1937 *Pictorial Plan of Glasgow* prepared by Leslie Bullock (1904–71) for Bartholomew & Son. Bullock had a close association with the Edinburgh publisher and this depiction of the city appears to be his first work. Born in London, he entered the Civil Service in 1920 and was trans-ferred to Scotland in 1932. He was involved in Civil Defence both during and after the Second World War, being awarded an O.B.E. in 1948. A fondness for children led to his *Children's Book of London*, published in 1939 and illustrated by a map which was originally sold in aid of Great Ormond Street Hospital. In the following years, he was to produce a series of historical maps of England and Wales, Ireland and Scotland which ran to several editions.

Like many of its immediate contemporaries, it indicates the railway lines and major roads running out of the city, as well as their destinations. Major public buildings, works, schools, hospitals, parks and sports venues, including two greyhound racecourses, are shown but, unlike earlier exhibition memorabilia, the Glasgow Subway was not identified. By far the most distinctive features of the map are the names, flags and coats of arms of the participant parts of the Empire. Pride of place naturally goes to the City's own insignia but those who study the map may be confused by the presence of the Star of India. Like the Exhibition itself, many of the names around the margins have passed into history but its whimsical cartoon style surely reflects the fun of the whole event.

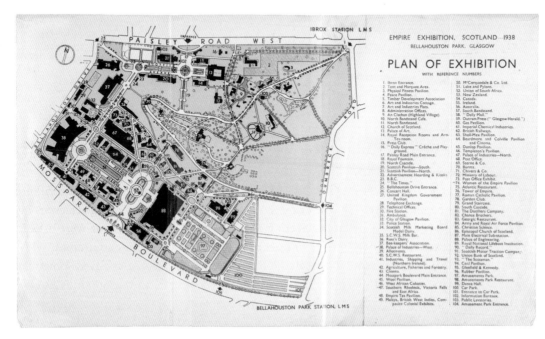

Plan of Exhibition with reference numbers (1938)

1938

War clouds gather: the Ordnance Survey prepares

On 3rd September 1939, Britain declared war on Germany but preparations had been in progress for at least a year before hostilities began. The government expected major air attacks on all British cities and, with the realisation that another major European conflict would disrupt and threaten the lives of those left at home, a variety of precautions were taken. These included the recruitment of civilians into a range of essential positions such as Air Raid Wardens and the Home Guard. By the time of the Munich crisis in September 1938, half a million citizens had volunteered for duties as wardens alone. To support the planning for national defence, particularly for air raid precaution (ARP), the Ordnance Survey issued a 'Special Emergency Edition' of the six-inch County series quarter sheets in 1938–39. It appears to have been restricted to urban areas in Britain where the population was in excess of 2,000 and the neighbouring counties to Glasgow were all scheduled to be printed in July 1939. All the six-inch sheets covering the city itself were based on earlier editions from 1914 or 1932–33.

Most of the updating relates to new buildings and roads, and was obtained variously from recent revisions at other scales and special 'ad hoc' surveys. As the official title implies, it was outside the normal programme of national mapping and was done hurriedly. This can be seen in the way new streets are drawn. Many of the rows of recently built semi-detached houses are indicated as continuous terraces without infill. Although these maps are not as attractive as other Ordnance Survey series, they are of great value as a record of how far our urban areas had expanded on the eve of the Second World War. In addition, as no copies were sent to the legal deposit libraries and they were very much used as working documents in a time of war, they are very rare.

In comparison with the contemporary Bartholomew *Plan of the County of the City of Glasgow*, prepared for the *Post Office Directory* of 1939–40, the level and amount of detail

Ordnance Survey, *Lanarkshire. Sheet 6 SE. Special Emergency Edition* (1938)

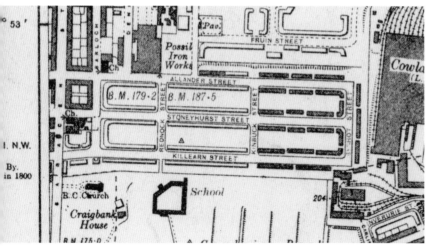

An example of the detail of new housing added to this edition; in this case, the streets south of the Possil Iron Works

which is immediately noticeable on the Special Emergency Edition underlines its value to an understanding of the city at the outbreak of war. By the use of different grades of shading, important factories, works, hospitals and other institutions are highlighted in such a fashion as to be instantly recognisable. The two selected map examples show the eastern half of Glasgow, covering the important railway works at Cowlairs and St Rollox, the Eclipse steel mills, the Saracen foundry, Robroyston colliery, Parkhead Forge, the Govan and Blochairn iron works, the Phoenix Tube Works at Dalmarnock and the Monkland Canal. They highlight how much heavy industry was located within or near the city itself. New houses and streets can be seen north of Cowlairs Park, off Springfield Road and at Provanmill. While all of these features are marked in close detail, only an outline of Barlinnie Prison is provided.

The commercially available Post Office plans of Glasgow were regularly updated by Bartholomew but the degree of detail and information could be variable. A consideration of similar coverage in the east of the city highlights the lack of depiction of such features as Templeton's carpet factory, Garngad brickworks and quarry, many of the works to the west of Carntyne station, Provan chemical works and the Kelvin engineering works on the directory map. Conversely,

the Bartholomew identification of theatres, cinemas, public baths and halls, and libraries is better and does include individual buildings at Barlinnie Prison. The new roads which had been added hastily to the Ordnance Survey sheets also appear on the *Directory* maps but, as their depiction is limited to lines only, there is no sense of building shape and identifying them is less straightforward.

Each sheet in the Emergency Edition carried a note under the title indicating that it was 'not on general sale to the public' and, while it might be thought that it was fortunate that these were not widely available, the reality is that the wealth of information on Glasgow's industries was already there from the earlier Ordnance Survey editions. The restriction on availability seems less to do with security and more with the purpose behind the production of the map series. It is known that enemy agents gathered relevant military information from a wide range of sources but a careful comparison of these two contemporary illustrations suggests that it was the Bartholomew map which was prepared with a greater degree of economy with the truth.

Despite the Depression of the early 1930s, Glasgow's population had reached a peak of more than 1,128,000 by 1939. This figure swelled to 1.3 million with the need for an increased workforce in the engineering plants and armaments factories in and around the city. With better employment opportunities and the immediate threat of attack still distant, life in Glasgow in the summer of 1939 was virtually normal. However, in preparation for the city's defence, the first blackouts were imposed and the volunteer ARP wardens patrolled the streets to ensure that buildings exposed no light as a guide for enemy aircraft. With typically resilient Glasgow humour, Dave Willis, one of the leading music-hall comedians of the day, created the character of 'the nicest looking warden in the ARP' and entertained audiences with his best loved song 'In my wee gas mask'.

OPPOSITE. The contemporary plan of the city by Bartholomew shows greater detail of public buildings (e.g. Barlinnie Prison, Rex Picture House) but less for the industrial works west of Carntyne Station

1941

A German military map of the city

All military strategy is based on good intelligence and this includes the gathering of maps and other cartographic information to help identify the location of the enemy and its sources of supply. It was only with the development of powered flight and the possibilities of aerial bombardment that large numbers of civilians distant from any conflict's front line might be subject to, and experience, attacks. During the First World War, London in particular had been the target for several Zeppelin raids but the main outcome of these was the creation of the Royal Air Force in 1918 and the system of early warning defence used so effectively in the Battle of Britain. Contrary to the generally accepted view, the German *Luftwaffe* rejected the theory of 'terror bombing' as it was seen as counter-productive and a diversion from the main aim of destroying the enemy's military forces. Nevertheless, the experience of German bombing raids on Guernica during the Spanish Civil War and on Rotterdam and Warsaw in the early months of the Second World War underlined the expectations of the British government of major air attacks.

The invasion and occupation of Norway, Denmark and France brought targets on the west coast of Scotland within range of German bombers. In particular, the Clyde ship-building yards and munitions factories, such as the Rolls Royce aero-engine plant at Hillington, put the whole area under the threat of air bombardment. *Luftwaffe* pilots were provided with clear aerial reconnaissance photography to help identify specific industrial targets. While London was to experience sustained bombing from late 1940 onwards, the Clyde area in general saw only a small number of lone raiders dropping comparatively few bombs. The city suffered its first raid in July 1940, when three people were killed, and a sequence of other attacks struck locations in the city centre (including the fruit market in Candleriggs), Maryhill, Dennistoun, Carntyne and Easterhouse.

Generalstab des Heeres, *Stadtplan von Glasgow mit Mil.-Geo.Eintragungen* (1941)

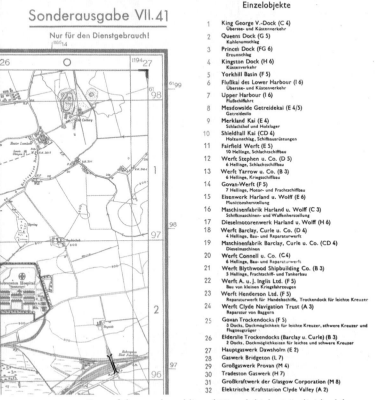

Einzelobjekte

1 King George V.-Dock (C 4)
Übersee- und Küstenverkehr
2 Queens Dock (G 5)
Kohlenumschlag
3 Princes Dock (FG 6)
Erzumschlag
4 Kingston Dock (H 6)
Küstenverkehr
5 Yorkhill Basin (F 5)
6 Flußkai des Lower Harbour (I 6)
Übersee- und Küstenverkehr
7 Upper Harbour (I 6)
Flußschiffahrt
8 Meadowside Getreidekai (E 4/5)
Getreidesilo
9 Merkland Kai (E 4)
Schlachthof und Holzlager
10 Shieldhall Kai (CD 4)
Holzumschlag, Schiffsausrüstungen
11 Fairfield Werft (E 5)
10 Hellinge, Schlachtschiffbau
12 Werft Stephen u. Co. (D 5)
6 Hellinge, Schlachtschiffbau
13 Werft Yarrow u. Co. (B 3)
6 Hellinge, Kriegsschiffbau
14 Govan-Werft (F 5)
7 Hellinge, Motor- und Frachtschiffbau
15 Eisenwerk Harland u. Wolff (E 6)
Munitionsherstellung
16 Maschinenfabrik Harland u. Wolff (C 3)
Schiffsmaschinen- und Waffenherstellung
17 Dieselmotorenwerk Harland u. Wolff (H 6)
18 Werft Barclay, Curle u. Co. (D 4)
4 Hellinge, Bau- und Reparaturwerft
19 Maschinenfabrik Barclay, Curle u. Co. (CD 4)
Dieselmaschinen
20 Werft Connell u. Co. (C4)
6 Hellinge, Bau- und Reparaturwerft
21 Werft Blythwood Shipbuilding Co. (B 3)
5 Hellinge, Frachtschiff- und Tankerbau
22 Werft A. u. J. Inglis Ltd. (F 5)
Bau von kleinen Kriegsfahrzeugen
23 Werft Henderson Ltd. (F 5)
Reparaturwerft für Handelsschiffe, Trockendock für leichte Kreuzer
24 Werft Clyde Navigation Trust (A 3)
Reparatur von Baggern
25 Govan Trockendocks (F 5)
3 Docks, Dockmöglichkeit für leichte Kreuzer, schwere Kreuzer und Flugzeugträger
26 Elderslie Trockendocks (Barclay u. Curle) (B 3)
2 Docks, Dockmöglichkeiten für leichte und schwere Kreuzer
27 Hauptgaswerk Dawsholm (E 2)
28 Gaswerk Bridgeton (L 7)
29 Großgaswerk Provan (M 4)
30 Tradeston Gaswerk (H 7)
31 Großkraftwerk der Glasgow Corporation (M 8)
32 Elektrische Kraftstation Clyde Valley (A 2)

Part of the numbered list of 'Einzelobjekte' (individual features) pinpointing docks, shipyards, gas works and important public buildings to be targeted by invading troops

All this changed on the nights of 13th and 14th March 1941 with two devastating raids by more than 200 bombers on the Clydebank shipyards. People who experienced the first night described Central Scotland as being lit by a 'bomber's moon'. Although the German strategy failed to damage local industry to any extent, the loss of life and the degree of destruction to the burgh's housing stock made it the most memorable Scottish raid of the war. As many bombs fell on Glasgow itself during the same attack, particularly in Knightswood, Partick and Drumchapel, including a direct hit on Bankhead School which killed 46 people and a mine explosion in Tradeston so powerful that it killed three sailors in the Broomielaw on the other side of the river. At the end of these two raids, 647 people

had died in the city with 1,680 others left injured.

While these figures are put into stark perspective by the high loss of life resulting from Allied bombing of the German cities of Hamburg and Dresden, they remind us that the war on the Home Front could be just as deadly as elsewhere. Later raids across the whole Clyde basin occurred during the following two months, but with the invasion of the Soviet Union in June 1941 the *Luftwaffe*'s attention was diverted elsewhere. However, not until February 1942 were forces detailed for an invasion of Britain released to other duties.

Prior to 1937, the German General Staff (*Generalstab des Heeres*) had been banned from collecting intelligence on Great Britain and its efforts to gather relevant information was focussed into the period from then until October 1940, when invasion plans were abandoned. The responsibility for the production of maps for use by the military was in the hands of the *Abteilung für Kriegskarten und Vermessungswesen* and it produced Special Editions (*Sonderausgaben*) of the relevant country's national mapping but with a German grid. The maps were dated, and in this example the sheet was prepared after corrections by aerial reconnaissance two months earlier, in July 1941 – in other words, after the immediate threat of invasion had passed. Nevertheless, this map was prepared with a military purpose, as denoted by the sub-title '*mit Mil.-Geo.-Eintragungen*', suggesting that the *Wehrmacht* still had a requirement for such mapping.

The marginal information indicates that the plan was based on the Ordnance Survey six-inch maps and these have been reduced to a scale of 1:15,000. The wealth of detail on the city's industries and other important elements, as mentioned on the Special Emergency sheets (1938), has been categorised by a key of coloured symbols which have been overprinted on the map. In fact, the addition of colour to indicate water bodies and parks enhances the original depiction. Thus highlighted are main roads, railways, bridges, docks, specific types of works, hospitals, barracks and important public utilities. Additionally, a supplementary list of 129 individually numbered targets (*einzelobjekte*), along with additional works not given a precise location, emphasises

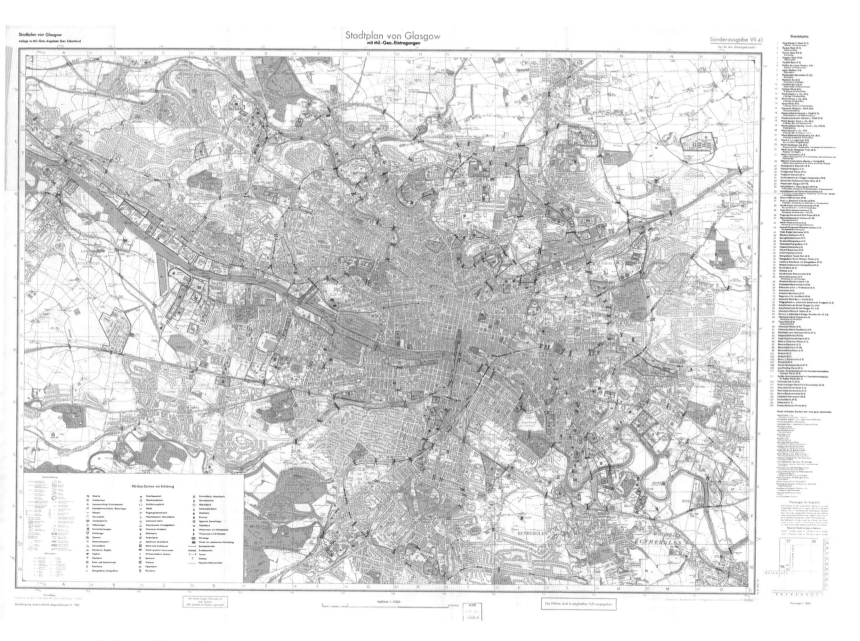

the key buildings of interest to military strategists.

Fascinatingly, the additional housing and street detail recorded on the Special Emergency maps is replicated and extended on this plan, reflecting a thoroughness in the German ability to gather comprehensive intelligence. In the areas of Cowlairs Park and Provanmill, the new housing is more detailed and more definite, while the prison buildings at Barlinnie are clearly shown. Given the size of the map sheet and the complexity of detail, it is doubtful if it was of use to German bomber pilots. However, had the German armed forces made a successful invasion, this would have been a very important document in any battle for Glasgow.

FIRST PLANNING REPORT
INNER CORE OF THE CITY

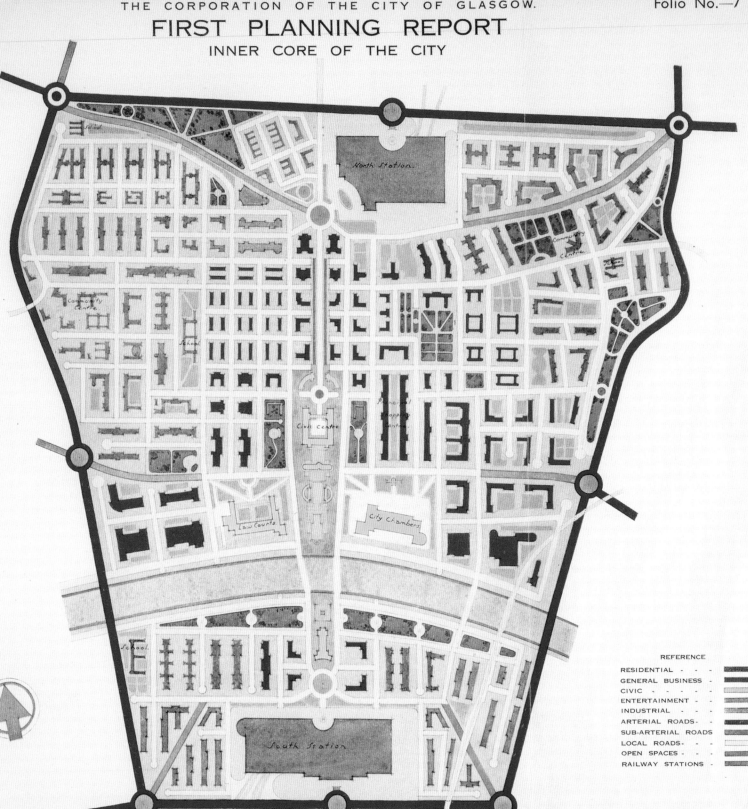

REFERENCE

RESIDENTIAL - - -
GENERAL BUSINESS -
CIVIC - - - -
ENTERTAINMENT - -
INDUSTRIAL - - -
ARTERIAL ROADS- -
SUB-ARTERIAL ROADS
LOCAL ROADS - - -
OPEN SPACES - - -
RAILWAY STATIONS -

The lines of the arterial and sub-arterial roads shown on this Plan are
in accordance with the road proposals for the City (see Folio No. 1).
The manner of development is in accordance with that shown on Folio No. 6.

ROBERT BRUCE, B.Sc., M.Inst.C.E., M.Inst.M. & Cy.E.
Master of Works and City Engineer,
OFFICE OF PUBLIC WORKS, CITY CHAMBERS, GLASGOW.

SCALE OF FEET
100 0   500   1000   2000   3000

1945

Post-war vision of a new
Glasgow: the Bruce plan

Compared with many other European cities, Glasgow's urban structure survived the Second World War relatively unscathed. As discussed earlier (**1941**), a certain amount of damage had been inflicted by enemy bombing raids but, by the middle of 1941, these had virtually ceased. However, the pre-war social problems of overcrowding and poor-quality housing continued to exist. By 1942, the Corporation was already making plans for the regeneration of the city centre. The following year, William Kerr, Glasgow's Town Clerk, prepared a memorandum on the likely problems facing the city in its post-war planning. Before hostilities ended, the Master of Works and City Engineer Robert Bruce presented his *First Planning Report to the Highways and Planning Committee* in March 1945, accompanied by a portfolio of maps, plans and drawings. These striking images emphasise what were a radical, if not draconian, set of proposals which would have completely changed the layout of the inner city. They are taken

from the portfolio and highlight the degree to which the Bruce Plan (as it is frequently named) sought to tackle the inherited problems faced by the authorities.

There were considerable tensions behind the various proposals being suggested and the city authorities were equally concerned about any loss of political power and autonomy if large numbers of its inhabitants were moved to locations well beyond the city boundary, as was suggested by Sir Patrick Abercrombie's *Clyde Valley Regional Plan* of 1949. Some of these issues continue to affect the whole Clydeside area. Bruce's report proposed a rebuilding of the city's central business district on a vast scale. This would have resulted in the demolition of such notable buildings as the City Chambers, Central Station and the Mackintosh-designed School of Art, as well as many of the other properties which today constitute the Merchant City. In addition, new higher-standard inner-city housing would be built to encourage a return of the wealthier

Robert Bruce, *First Planning Report to the Highways and Planning Committee. Inner Core of the City. Folio 7* (1945)

Bruce's vision of a new Civic centre still retaining a geometric pattern

the overcrowded centre would be swept away. There was a strong, almost evangelical, spirit of reconstruction in the late 1940s, which, in Glasgow, was heavily influenced by the Modernist style of architecture. As the plans show, this was characterised by a markedly geometric approach to urban design, with regularly spaced tower blocks built in functional districts and ringed by a motorway network. While many see the influence of the French architect Le Corbusier in the portfolio put forward by Bruce, there is equally a similarity with several of Albert Speer's monumental designs for Hitler's Germany.

Although the concept was quite staggering in its scale, the city's Planning Committee initially approved the plan in 1947. Two years later, an exhibition entitled 'Glasgow: today and tomorrow' opened in the Kelvin Hall, designed to display the new design of Bruce's concept and a film was commissioned by the civic authorities to be shown in cinemas, schools and other community venues. However, by that date, a combination of concern, economics and political influence resulted in the plan being dropped.

Despite the authorities rejecting the more extreme ideas in reconstructing Glasgow, Bruce's suggestion of decanting considerable numbers of citizens to satellite townships within the city boundary was to influence subsequent planning, leading to the creation of large housing estates in such areas as Drumchapel, Easterhouse and Castlemilk. It also led to the creation of the several Comprehensive Development Area schemes which lasted until the late 1970s. Often characterised by incomplete execution and budgetary constraints, the debate on the impact of such plans as these on the ordinary citizen and the communities involved continues.

Fundamental to the plan was a co-ordinated road transport system, supported by two new railway hubs on either side of the Clyde linked by a central axis. It took the Beeching railway reports of the mid-1960s to rationalise the local railway network to something akin to this original concept. Bruce left his Engineer's post before his plan was discarded and he died in a car crash in 1956. The vision he had for a motorway ring-road is, however, one element which

residents to the core area, while those less well off would be moved to the periphery. Industry would also be removed from the centre. It reflected optimism in modern technology as a solution to many of the problems. Bruce was under no illusion about the length of time all this would take. He believed that a city as important as Glasgow could not be treated in a piecemeal fashion and the plan would require 50 years to achieve its goals.

It is difficult to discuss this somewhat futuristic vision for Glasgow without using emotive language and many subsequent writers have condemned aspects of the ideas as 'social engineering'. Overall, it envisaged an ideal, ordered and efficient city in which the decayed and squalid conditions of

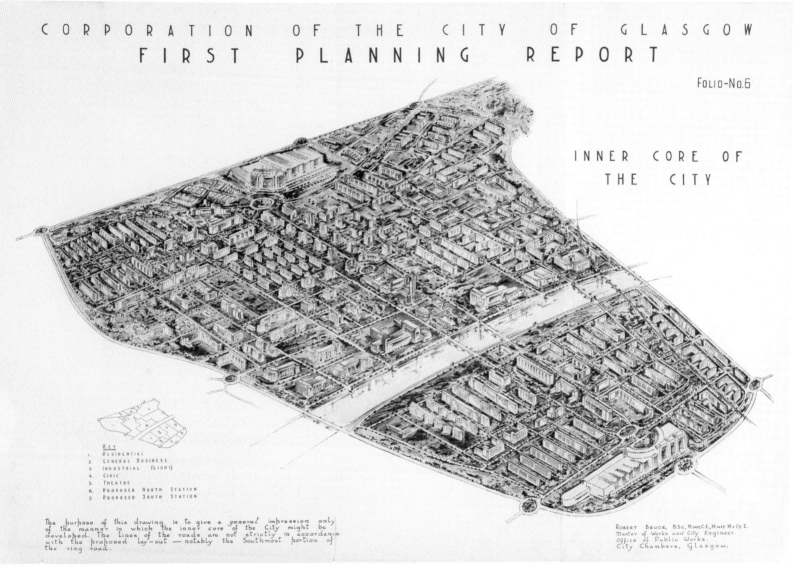

CORPORATION OF THE CITY OF GLASGOW
FIRST PLANNING REPORT

FOLIO-No.6

INNER CORE OF
THE CITY

KEY
1. RESIDENTIAL
2. GENERAL BUSINESS
3. INDUSTRIAL (LIGHT)
4. CIVIC
5. THEATRE
6. PROPOSED NORTH STATION
7. PROPOSED SOUTH STATION

The purpose of this drawing is to give a general impression only of the manner in which the inner core of the City might be developed. The lines of the roads are not strictly in accordance with the proposed lay-out — notably the Southmost portion of the ring road.

ROBERT BRUCE, B.Sc, MInstC.E., MInst M. Cy. E.
Master of Works and City Engineer.
Office of Public Works.
City Chambers, Glasgow.

Bruce may have been influenced by Albert Speer, but the dominant buildings to the north and south are proposed railway stations

was realised, albeit in a fragmented fashion.

The 'look' of any city reflects the past decisions made by those who administer it and its residents. In turn, they are influenced by the mood and attitudes of their day. It could be argued that the architectural heritage which is celebrated in Glasgow today is a treasure which has been saved not from the ravages of war but from the beliefs of a new era. Intriguingly, the back cover of the Bruce portfolio is illustrated by Barry's 1782 city plan, which, if nothing else, may show his awareness of the city's history and belief in a progressive future.

GLASGO

1948-50

Air-photo mosaics: a short-term mapping solution

Although the Ordnance Survey had earlier experimented with air photography in the revision of mapping based on ground survey, it was only with the urgent need for post-war planning that series of photographic mosaics were produced. Much of the detailed topographic mapping from the 1930s was considerably dated by then and the scale of the demand required an ambitious but speedy solution. With agreement from the Air Ministry, flight surveys of areas of priority were undertaken by the RAF between 1944 and 1950, based on advice from the Ministry of Town and Country Planning. There was a particular need for up-to-date mapping for cities badly damaged during hostilities. Given this stated emphasis on the need for reconstruction, it is a little surprising, from a Scottish perspective, that there is no coverage at all for Clydebank and only 221 mosaics of Scotland were published during this programme.

The resulting photo-maps were produced originally at 1:1250 and, under pressure from planning authorities, at 1:10,560, based on newly designed National Grid sheetlines. The latter series covers a much wider area and is the better known. Scotland does not appear to have had any maps prepared at the more detailed scale, which was used mostly for London, Swansea and several south-coast towns. By late 1947, however, there was extensive coverage at the six-inch scale for central Scotland and South Wales. An investigation of the Scottish mosaics identifies two major concentrations of images. The Firth of Forth has a considerable swathe of photographs for both sides of the estuary from its mouth to Kincardine and Stirling. It is significant that there is no coverage for the naval base at Rosyth. In the west, the Clyde Valley was mapped from Lanark to Glasgow but is noticeably incomplete, particularly for parts of the city. A further group of images focuses on the Ayrshire coast inland to New Cumnock. Elsewhere, individual towns and cities, such as Biggar,

Ordnance Survey, *Air Photo Mosaic NS66NW* (1946)

Galashiels, Fort William and St Andrews were photographed.

Each photograph was enlarged, rectified to compensate for distortion and fitted into a mosaic of images. Place names and borders were added, the maps being compiled by a specialist division of the Ordnance Survey. While the original intention had been to prepare these for government departments, they were put on general sale in order to recoup some of the production costs. Publication of the mosaics began in 1946 but, when the proposal to sell them to the public was taken, the issue of national security became of growing concern. Although the Ordnance Survey had a definite policy about the omission of details of highly sensitive military and other important installations *prior* to production of their maps, the mosaics required to be edited *after* being completed. In the era of the Cold War, this was of major significance and resulted in the introduction of security checks, combined with the intentional misrepresentation of key defence features, either by adding cloud cover, often using cotton wool, or by replacing sites with patchworks of fields or woods. For some areas, the required revisions were so great that the sheets were withdrawn completely and, in most cases, the urban coverage was never fully finished.

Since there was only partial coverage of the city, neither of the mosaics illustrated shows the centre of Glasgow. They have been selected to emphasise different aspects of this series and are recorded, rather prosaically, as sheets NS66NW (Lanarkshire) and NS46 NE (Renfrewshire). The first image covers part of the area shown on the Special Emergency Edition (1938) maps of the eastern fringes of Glasgow. Clearly to be seen on the photo-map are the Cowlairs and St Rollox railway works, the buildings within Barlinnie Prison, the spoil heap at Robroyston Colliery and the Monkland Canal, as well as much of the new housing which had first appeared on the earlier map. While the Renfrewshire map falls outside the city's boundary, it is significant in showing the runway of Renfrew Airport, which served as the city's domestic airport until 1966 and which had also been used by the RAF during the Second World War. Clearly, no one thought that leaving the depiction on the image was a threat to national defence!

As a result of the high sale price for the individual maps and the lack of comprehensive sets for many cities, they were never popular with the public. In addition, the Ordnance Survey appeared to be reticent about their promotion and the Treasury was so concerned by the low demand that it wanted the programme curtailed. By 1952, the mosaics were slowly being replaced by more conventional mapping and sales were suspended in early 1954. Certainly, there seems to have been an ambivalent attitude by senior staff within the Ordnance Survey and what had been seen as an important, but temporary short-term solution became the victim of financial stringency.

Despite the incompleteness of the series, the detail in the photographs is far greater than that which is available from other contemporary mapping. Those mosaics which were published are particularly valuable in helping to identify contaminated land and earlier land uses for developers. Unlike paper maps, there is no generalisation of features and the mosaics represent the first use of aerial survey methods across a wide area. The Ordnance Survey itself did not use them to revise any subsequent mapping as they were not considered sufficiently accurate. Modern mapping continues to be based largely on aerial photography but there is still a need for staff to visit and review areas that cannot be surveyed by photogrammetric methods. An example of this would be land obscured by vegetation. Current policy is to aim to ensure that major features (such as new motorways or large housing developments) are surveyed within six months of construction, using GPS (global positioning system) technology to guarantee high standards of precision.

OPPOSITE. This example from Renfrewshire shows the runway at Renfrew Airport, the city's domestic airport until 1966

MOORPARK

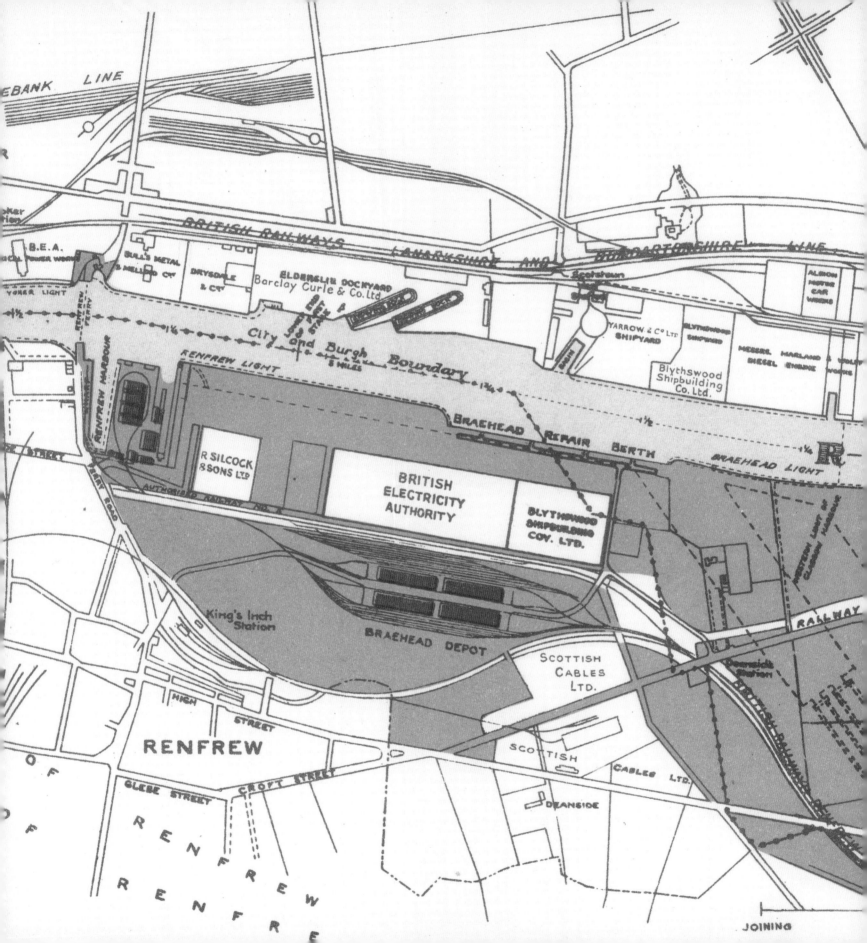

1955

The Clyde at the end of its heyday

The traditional Scottish Hogmanay custom of opening windows to let the old year out and the New Year in was enhanced in Glasgow by ships in harbour sounding their sirens to herald the event. When this map was prepared, Glasgow was still a bustling and busy harbour, with a degree of zoning of vessels. By this date, the upper harbour quays were used mainly for barges unloading sand and gravel, while the lower quays continued to be allocated to coastal shipping. Change was already in progress. By 1954–55, coal exports from the General Terminus Quay had shrunk to less than 10% of the tonnage shipped at the beginning of the century and new facilities for the unloading of iron ore were being planned. Downriver, recognising the increasing size of vessels, the King George V Dock at Shieldhall had been opened in 1931 to accommodate ocean-going vessels and, on the north bank, the Clyde Navigation Trust had built the massive Meadowside Granary, the Merklands Animal Lairage and its neighbouring fruit-sheds. By 1956, the linear length of quay space amounted to more than 19.7 kilometres.

A cruise on the river between the Broomielaw and Clydebank would have passed the shipyards of many famous builders, including Harland and Wolff, Fairfield, Barclay Curle, Yarrow, and Lobnitz. Between 1946 and 1948, Britain built and launched just over half the world tonnage of merchant vessels, with over 50 per cent coming from the Clyde shipyards alone. As late as 1967, the *Queen Elizabeth II* was launched as the flagship of the Cunard Steamship Company from the John Brown Clydebank yard. However, the writing was already on the wall for many businesses as overseas competition grew and the decline in transatlantic passenger numbers saw the *QE2* latterly operate more as a cruise ship.

This plan originally appeared eight years earlier, dated 1945, as one of a series of insets included in an extensive guide to the port of Glasgow published by the Clyde Navigation

Archibald Thomson, *Clyde Navigation Trust. River Clyde from Glasgow to Clydebank* (1955)

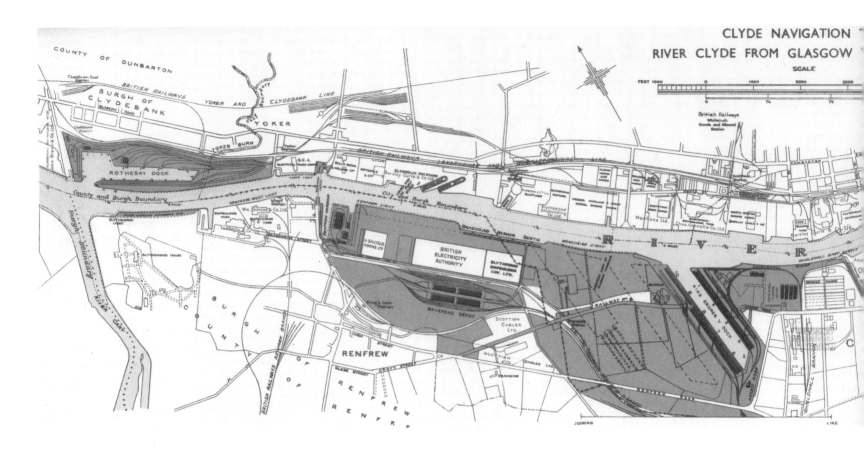

Trust, who described it as 'one of the world's great ports'. Its Chairman, James Leggat, avoiding any comment on current trends elsewhere, expressed good hopes for its future, based on hard and steady work. The text provided details of each quay and basin, along with information on the berths and cranes. By this date, over 100 shipping companies either traded with, or were registered in, the port. With hindsight, it is clear to see in the whole volume an assuredness in the strength of the extensive and complex industrial basis of this success, as exemplified by the sheer number of businesses advertisements (the list alone running to eleven pages). It may also possibly reflect a lack of awareness of the many changes which the post-war era ushered in.

Both plans were prepared by Archibald Thomson, engineer to the Trust from 1945 until 1962, and much of the detail is adapted from the earlier work of James Deas (1898–

99). Berths are numbered and the complex juxtaposition of quays, docks, shipyards and sheds emphasises the narrowness of not only the river but the harbour area itself. The considerable tract of undeveloped land at Braehead purchased by the Trust downstream of the King George V Dock reflects an optimism which never came to fruition.

In 1958, only three years after this plan was published, the British Association met in Glasgow for the sixth time. In preparation, Ronald Miller and Joy Tivy, both members of the University's Geography Department, edited *The Glasgow Region: a general survey*. Its foreword may reflect a degree of uncertainty in its almost apologetic claim that 'visitors are likely to know little of us: Glasgow does not rank as a tourist attraction'. More significantly, its review of industry and commerce highlighted the underlying problems faced by a river basin which was not well placed for the transition from

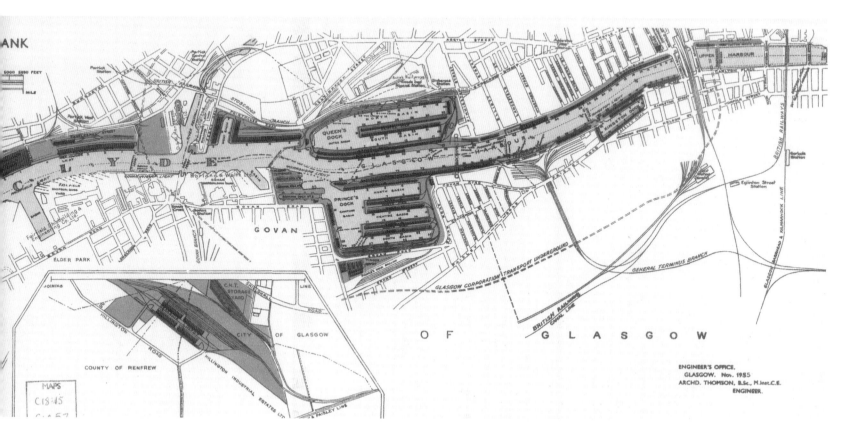

an overseas to a UK market. As was subsequently stressed, Glasgow 'twenty miles up a narrow river' could not compete with the facilities offered at such docks as Southampton, Liverpool and Tilbury, all better placed to serve the major British centres of population.

As a port, Glasgow accounted for only about 4 per cent of the British overseas trade and continued to function more as a regional centre, with imports accounting for 70 per cent of the volume total, largely made up of grain and other foodstuffs and petroleum. Passenger sailings continued but were largely restricted to local services by pleasure steamers, as well as connections to the west coast and Ireland. Regardless of this contraction of business, the head offices of some thirty companies were based in the city, along with insurance and other ancillary services.

By the middle of the twentieth century, improvements in transport technology began to have a serious impact on shipping on the upper Clyde. The growth in containerisation of goods reduced the demand for ships and, combined with the increase tonnage of vessels, made use of the upper harbour quite impractical. In addition, competition from road haulage, private car ownership and cheap air travel to sunnier holiday destinations hastened the dramatic decline of the port and shipyard facilities in Glasgow. At the end of 1965, the Navigation Trust was dissolved in a merger with other bodies to create the Clyde Port Authority, twenty years after this had been proposed by a government committee chaired by Lord Cooper. Five years after this consolidation, Queen's Dock was closed to shipping. As one notable historian of the river, John Riddell, has stated, 'the Clyde is in the wrong place in the days of motorways, Europe and speed'.

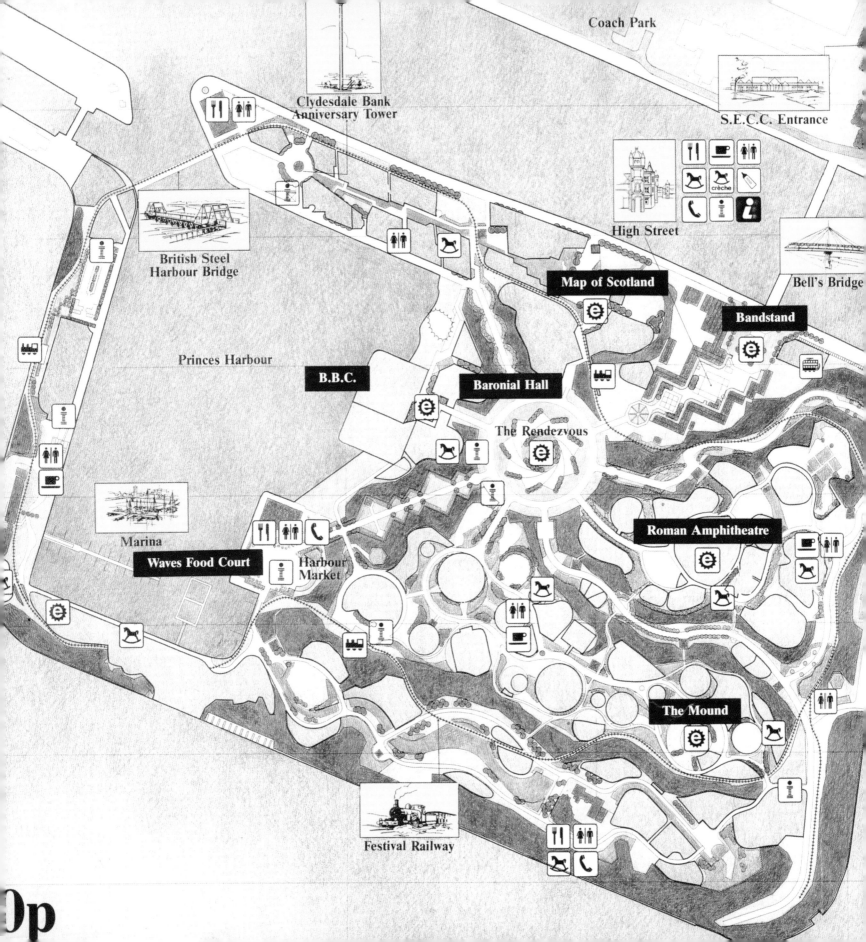

Coach Park

S.E.C.C. Entrance

Clydesdale Bank
Anniversary Tower

High Street

British Steel
Harbour Bridge

Bell's Bridge

Map of Scotland

Princes Harbour

Bandstand

B.B.C.

Baronial Hall

The Rendezvous

Roman Amphitheatre

Marina

Waves Food Court

Harbour
Market

The Mound

Festival Railway

0p

1988

Glasgow's Garden Festival: *a sign of regeneration*

Like many other industrial regions within the United Kingdom, Glasgow's reliance on heavy engineering continued well into the second half of the twentieth century, despite the warnings and experiences of the inter-war period. Writing in the *Third Statistical Account* for Glasgow in 1958, R.H. Campbell remarked that the degree of dependence on such industries was a dangerous feature of trade in the city but he voiced the hope that it might continue in prosperity, if quite limited in the possibilities of expansion. Growing overseas competition, declining demand, a lack of managerial vision, under-investment and inflexible labour attitudes all contributed to falling production, increasing unemployment and economic recession.

In spite of considerable government intervention and subsidy, many of the major industries in which Glasgow had led the world and taken great pride began to disappear slowly. This decline was not uncontested and such events as the 1971 Upper Clyde Shipbuilders work-in aimed to dispel the myth of a 'work-shy' labour force, as well as underlining both the right to work and the long-term viability of the yards. Although it would be an exaggeration to say that the city lost confidence, parts of Glasgow became desperate places, experiencing a lack of local employment opportunities and particularly affected by some of the more severe free-market policy decisions of the Conservative governments of the 1970s and 1980s.

The garden festival concept began in Germany following the end of the Second World War as part of the redevelopment of derelict land, with the combined intentions of returning such areas to industrial, commercial, recreational or housing use and of bringing people together. Although the concept was imported to Britain, the five festivals held between 1984 and 1992 were more a part of a wider government strategy aimed at involving the private sector in inner-city regeneration and

Glasgow Garden Festival '88 (1988)

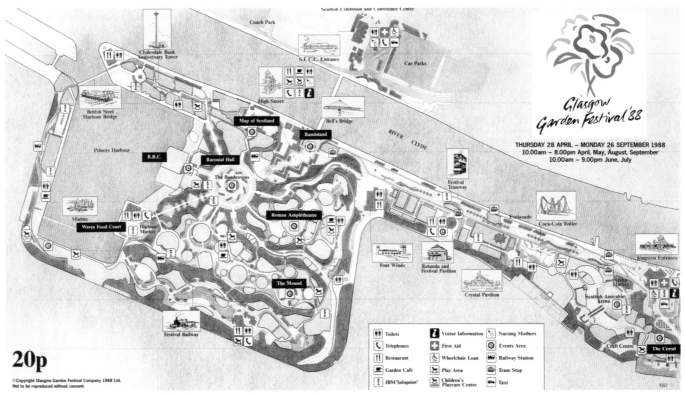

The complete Festival plan shows how compact the actual Garden area was

had little to do with promoting horticulture. In consequence, the whole programme had a markedly weaker record in post-event follow-through than other parts of Europe and was subsequently criticised for being either too commercial, lacking cohesive vision or failing to create long-term public urban parkland.

While many may see the idea of holding these events in some of the nation's more depressed industrial areas as something of a temporary distraction rather than a conscious effort by central government to tackle the major issues faced in them, it is now recognised that the Glasgow experience was another step on the road towards the city becoming European City of Culture in 1990. Three years before the Festival, the Scottish Exhibition and Conference Centre was opened on the site of the former Queen's Dock as the premier national venue

for public events, concerts and major expositions. Part of its main building was used for shows at the Garden Festival and the construction of Bell's Bridge helped tie the venue sites together.

Regardless of later discussions of economic and social benefits, there is no question that the Glasgow event was a resounding and popular success. The 4.3 million visitors who attended far exceeded both local expectations and the figures at any of the other four festivals. Certainly, much of the success was due to strong local support (season ticket sales numbered over 100,000), substantial sponsorship and to the overall control of the venture by the Scottish Development Agency, whose Chief Executive, George Mathewson, chaired the separate company set up to run the event. As an exercise in image building and promotion, it played a significant role in helping increase Glasgow's contemporary popularity with many visitors and was yet another example of its ability to re-invent itself.

Much of the promotion of the Garden Festival benefitted from the successful 'Glasgow's Miles Better' campaign and had an equally commercial thrust, offering visitors a 'day out of this world'. With such elements as the Coca-Cola Roller and the return of a number of Glasgow trams as part of the entertainment, there were echoes of the sense of fun which had been a characteristic of the Empire Exhibition fifty years earlier. This aspect can also be seen to reflect the city's resilience and determination to plan for a better future.

As the illustration shows, the site for the Festival was the derelict Prince's Dock, which had the advantage of being non-toxic, easily cleared and partly filled in to increase the available area. At 64 hectares, it was notably smaller than its predecessors at Liverpool and Stoke but this helped focus visitor attention. A sense of busyness was one of the abiding memories for many who attended. Nonetheless, a major theme was also based on a return to nature, with a particularly strong appeal to children. The appointment of an educational co-ordinator led to almost every school pupil in the greater Glasgow area visiting at least once. One of the gardens, designed by James Cunning Young & Partners for the National Trust for Scotland and Dobbies, the Scottish nurserymen and garden centre owners, focussed on Great Scottish Plant Collectors and reflected different aspects of some of the Trust's own gardens, such as the herbaceous border at Falkland Palace.

A marina, established in part of the harbour, reflected an effort to reconnect the city with the river but it would be difficult to argue any subsequent success in this particular aim. Development of the area now known as Pacific Quay has seen both the BBC and STV relocate their headquarters and studios to the south bank. Along with Capital Radio station, they form a cultural precinct well situated for events at the SECC and the Hydro. On the other hand, the neighbouring graving docks still lie derelict and the Glasgow Tower has been plagued with problems ever since opening. While the profile of Glasgow's riverside heritage has been raised, the sight of the solitary paddle steamer *Waverley* berthed alongside the Glasgow Science Museum remains testimony to a need for greater promotion of the Clyde itself – a challenge Glasgow will undoubtedly tackle in its own way.

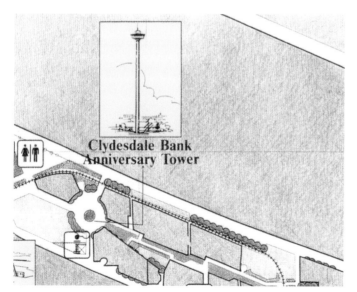 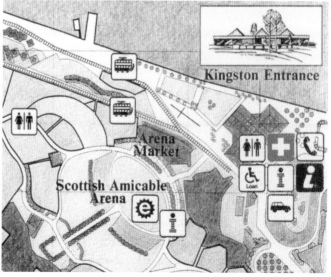

Two contrasting features of the Festival: the Clydesdale Bank Tower and the line of the very popular Festival Tramway

Epilogue

All Kinds of Folk: Folk of All Kinds. Hillhead 2011

The major underlying theme of the text in this book has been the relationship between individuals, the environment they live in and maps. This leitmotif is a constant, regardless of which images are selected. Another strand in the narrative has been Glasgow's strong links with art, education and transport. The symbiotic processes existing between a city, its residents and these other elements are brought together in this final image, created on ceramic tiles by the local artists Alasdair Gray and Nichol Wheatley. Unveiled in 2012, Gray's mural adorns the entrance to Hillhead Subway Station, the nearest to the University of Glasgow, and is an echo of some of the world's earliest cartographic depictions, whether carved on tablets or painted on walls.

The mural goes further. It is also a reminder of the artistry within maps and their decorative value. In a style which possibly alludes to the way in which Vermeer and other Dutch painters populated several interior compositions with examples of cartography, the artists have gifted a work of art for all to share. As with all of the maps in this text, Gray's vision is a snapshot in time and already documents a past geography. In this, it perhaps makes a stand against the present-day evolving technologies which affect maps in both their making and use. While today's cartography can appear more insubstantial, in the sense that any representation may easily be changed 'on demand', this picture of Hillhead is a tangible record which can be compared with past and future maps. Gray is as well known as a writer and, interestingly, his most celebrated work, *Lanark,* contains the postscript words 'I have grown up, My maps are out of date'.

More than a utilitarian record or decorative device, it is also a celebration of city life and the rich variety of talent shown within the chosen local community, whether 'bonny fighters', 'merry devils', 'bold explorers' or 'head cases'. Many of these flank either side of the central depiction of the Hillhead area.

Detail from Alasdair Gray, *Hillhead 2011* (2011)

Gray himself is recorded as commenting, 'I knew I would enjoy depicting it, and those who use the Subway, in a symbolic and humorous way.'

The 'happy combination' of imagery and expression which maps present will continue to have a hold upon the memory, if in a markedly different fashion from earlier times. Gray also gives us a valuable reminder of what is important in our lives:

'Do not let daily to-ing and fro-ing
To earn what we need to keep going
Prevent what you once felt when wee
Hopeful and free'

Further Reading

Cartography

Cunningham, I.C. (ed.), *The Nation Survey'd: essays on late sixteenth-century Scotland as depicted by Timothy Pont* (East Linton: Tuckwell Press, in association with the National Library of Scotland, 2001).

Fleet, C., Wilkes, M. and Withers, C.W.J., *Scotland: mapping the nation* (Edinburgh: Birlinn, in association with the National Library of Scotland, 2011).

Fleet, C. and MacCannell, D., *Edinburgh: mapping the city* (Edinburgh: Birlinn, in association with the National Library of Scotland, 2014).

Hewitt, R., *Map of a Nation: a biography of the Ordnance Survey* (London: Granta, 2010).

Hyde, R., *Gilded Scenes and Shining Prospects: panoramic views of British towns, 1575–1900* (New Haven: Yale Center for British Art, 1985).

Moir, D.G. (ed.), *The Early Maps of Scotland to 1850*, 2 volumes (Edinburgh: Royal Scottish Geographical Society, 1973 and 1983).

Moore, J.N., *The Historical Cartography of Scotland: a guide to the literature of Scottish maps and mapping prior to the Ordnance Survey* O'Dell Memorial Monograph No.24 (Aberdeen: University of Aberdeen, Department of Geography, 1991).

Oliver, R.R., *Ordnance Survey maps: a concise guide for historians*, 3rd edition (London: Charles Close Society, 2013).

Smith, D., *Victorian Maps of the British Isles* (London: Batsford, 1985).

Woodward, D. (ed.), *The History of Cartography, Vol.3: Cartography in the European Renaissance* (Chicago and London: University of Chicago Press, 2007).

Surveyors, engravers, printers

Bendall, S. (ed.), *Dictionary of Land Surveyors and Local Map-Makers of Great Britain and Ireland, 1530–1850*, 2nd edition (London: British Library, 1997).

Bushnell, G.H., *Scottish Engravers: a biographical dictionary … to the beginning of the nineteenth century* (London: Oxford University Press, 1949).

Schenck, D.H.J., *Dictionary of the Lithographic Printers of Scotland, 1820–1870: their locations, periods, and a guide to artistic lithographic printers* (Edinburgh: Edinburgh Bibliographical Society, 1999).

Worms, L. and Baynton-Williams, A., *British Map Engravers: a dictionary of engravers, lithographers and their principal employees to 1850* (London: Rare Book Society, 2011).

Glasgow

Bilsborough, P., The Development of Sport in Glasgow, 1850–1914. Unpublished M.Litt. thesis, University of Stirling, 1983.

Brown, J.A., 'The cartography of Glasgow with list of old maps and illustrations reproduced', *Scottish Geographical Magazine* 37 (1921), 67–75.

Crawford, R., *On Glasgow and Edinburgh* (London: Harvard University Press, 2013).

Curtis, E.W., *The Story of Glasgow's Botanic Gardens* (Glendaruel: Argyll Publishing, 2006).

Cunnison, J. and Gilfillan, J.B.S., *The Third Statistical Account of Scotland: Glasgow* (Glasgow: Collins, 1958).

Fraser, W.H., and Maver, I. (eds), *Glasgow. Vol.2 1830–1912* (Manchester: Manchester University Press, 1996).

Gibb, A., *Glasgow: the making of a city* (London: Croom Helm, 1983).

Kellett, J.R., 'Glasgow' in Lobel, M.D. (ed.), *Historic Towns: maps and plans of towns and cities in the British Isles, with historical commentaries, from earliest times to 1800. Vol.1* (London: Lovell Johns, 1969).

Kinchin, P. and Kinchin, J., *Glasgow's Great Exhibitions: 1888, 1901, 1911, 1938, 1988* (Bicester: White Cockade, 1988).

Maver, I., *Glasgow* (Edinburgh: Edinburgh University Press, 2000).

Moore, J.N., 'The Ordnance Survey 1:500 town plan of Glasgow: a study of large-scale mapping, departmental policy and local opinion', *Cartographic Journal* 32 (1995), 24–32.

Moore, J.N., *The Maps of Glasgow: a history and cartobibliography to 1865* (Glasgow: Glasgow University Library, 1996).

Moore, J.N., "Many years servant to the town': James Barrie and the eighteenth century mapping of Glasgow', *Scottish Geographical Magazine* 113 (1997), 105–112.

Niven, A.S., *Early Glasgow Maps: an index* (Glasgow: Mitchell Library, 1984).

Reed, P. (ed.), *Glasgow: the forming of the city*, 2nd edition (Edinburgh: Edinburgh University Press, 1999).

Riddell, J.F., *Clyde Navigation: a history of the development and deepening of the River Clyde* (Edinburgh: John Donald, 1979).

Walker, F.A., 'The Glasgow grid' in Markus, T.A. (ed.), *Order in space and society: architectural form and its context in the Scottish Enlightenment* (Edinburgh: Mainstream, 1982).

Williamson, E., Riches, A. and Higgs, M., *Glasgow: The Buildings of Scotland* (London: Penguin, 1990).

Withey, M., The Glasgow City Improvement Trust: an analysis of its genesis, impact and legacy and an inventory of its buildings, 1866–1910. Unpublished Ph.D. thesis, University of St Andrews, 2002.

Websites

Charting the Nation, http://www.chartingthenation.lib.ed.ac.uk. Contains over 3,500 images of early maps of Scotland, from 1550 to 1740.

Fire Insurance Maps and Plans, http://www.bl.uk/onlinegallery/onlineex/firemaps/fireinsurancemaps.html. The British Library's comprehensive holdings of fire insurance plans produced by the firm Charles E. Goad Ltd dating back to 1885, including images of each sheet which covers Glasgow.

The Glasgow Story, http://www.theglasgowstory.com. Covering all aspects of the city's history and including images of several maps and plans.

Map History/History of Cartography Gateway, http://www.maphistory.info. Includes a geographical listing of images of early maps.

National Library of Scotland – Map Images, http://maps.nls.uk. More than 20 maps of Glasgow are included on this site which provides access to over 100,000 maps of Scotland and beyond, c.1560–1960.

National Library of Scotland – Bartholomew Archive, http://digital.nls.uk/bartholomew. Background information, inventories and listings of this Edinburgh firm, one of the world's largest commercial cartographic archives.

OldMapsOnline, http://www.oldmapsonline.org. The world's largest portal to freely available historical maps, with a zoomable graphic bounding box.

Oxford Dictionary of National Biography Online, http://www.oxforddnb.com. Contains short referenced biographies of more than 59,000 British people of all periods, including a considerable number of cartographers, surveyors and publishers.

ScotlandsPlaces, http://www.scotlandsplaces.gov.uk. A collaborative portal to maps and other geographical resources, including those in the National Records of Scotland, National Library of Scotland and the Royal Commission on the Ancient and Historical Monuments of Scotland.

Index

Titles of books, maps and periodicals are arranged under their creators